Communications
in Computer and Information Science   1536

More information about this series at https://link.springer.com/bookseries/7899

Sang-Yoon Chang · Luis Bathen ·
Fabio Di Troia · Thomas H. Austin ·
Alex J. Nelson (Eds.)

Silicon Valley Cybersecurity Conference

Second Conference, SVCC 2021
San Jose, CA, USA, December 2–3, 2021
Revised Selected Papers

 Springer

Editors
Sang-Yoon Chang (ID)
University of Colorado
Colorado Springs, CO, USA

Luis Bathen
IBM Almaden Research Center
San Jose, CA, USA

Fabio Di Troia (ID)
San Jose State University
San Jose, CA, USA

Thomas H. Austin (ID)
San Jose State University
San Jose, CA, USA

Alex J. Nelson (ID)
National Institute of Standards
and Technology
Gaithersburg, MD, USA

ISSN 1865-0929   ISSN 1865-0937 (electronic)
Communications in Computer and Information Science
ISBN 978-3-030-96056-8   ISBN 978-3-030-96057-5 (eBook)
https://doi.org/10.1007/978-3-030-96057-5

This Springer imprint is published by the registered company Springer Nature Switzerland AG
The registered company address is: Gewerbestrasse 11, 6330 Cham, Switzerland

Preface

The 2nd Silicon Valley Cybersecurity Conference (SVCC 2021) took place virtually during December 2–3, 2021. SVCC facilitates research in dependability, reliability, and security to address cyber-attacks, vulnerabilities, faults, and errors in networks and systems. This conference provides a high-quality forum for participants to exchange their research in robustness and resilience in a wide spectrum of computing systems and networks. The conference addresses innovative system design, protocols, and algorithms for detecting and responding to malicious threats in dependable and secure systems and networks including experimentation and assessment.

SVCC 2021 featured five keynote speakers, from academia and industry, and six different research programs along with a distinguished research forum. The special research forum recognized three distinguished researchers in 2021 who presented their high-quality research in cybersecurity. The conference included ten research papers this year, with an additional session for poster presentation. Papers were evaluated with a double-blind review process, with three reviews per paper.

In addition, the conference had a special panel for Women-in-Cybersecurity, a Capture the Flag competition, and the UNiSEC Datathon. The goal of the Datathon was to create an exciting learning experience at the intersection of cybersecurity and machine learning. We were pleased to see a diverse range of participants, with 52.6% female students and 26.4% underrepresented students taking part in the Datathon challenge for one month.

We are grateful to the three conference sponsors for SVCC 2021: Cisco, Jabil, and Trend Micro.

December 2021

Sang-Yoon Chang
Luis Bathen
Fabio Di Troia

Organization

General Chairs

Divyesh Jadav — IBM Research, USA
Younghee Park — San Jose State University, USA

Program Chairs

Sang-Yoon Chang — University of Colorado, Colorado Springs, USA
Luis Bathen — IBM Research, USA
Fabio Di Troia — San Jose State University, USA

Publicity Chairs

Sara Tehranipoor — Santa Clara University, USA
Wei Yan — Clarkson University, USA

Publication Chairs

Thomas Austin — San Jose State University, USA
Alex J. Nelson — National Institute of Standards and Technology, USA

Registration Chairs

Michael Tjebben — L&T Technology Services Limited, USA
Sang-soo Lee — San Jose State University, USA

Poster Chair

Gokay Saldamli — San Jose State University, USA

Special Session Chairs

Nima Karimian — San Jose State University, USA
Hossein Sayadi — California State University, Long Beach, USA

Datathon Chair

Jorjeta Jetcheva   San Jose State University, USA

Technical Program Committee

Vikrant Nanda   Google Inc., USA
Subhash Lakshminarayana   University of Warwick, UK
Malek Ben Salem   Accenture Inc., USA
Sang Kil Cha   Korea Advanced Institute of Science and
                                   Technology, South Korea
Harshan Jagadeesh   Indian Institute of Technology, Delhi, India
Carlos Rubio-Medrano   Texas A&M University–Corpus Christi, USA
Eul Gyu Im   Hanyang University, South Korea
Lei Xu   University of Texas Rio Grande Valley, USA
Xiaoyan (Sherry) Sun   California State University, Sacramento, USA
Tai M. Chung   Sungkyunkwan University, South Korea
Liudong Xing   University of Massachusetts Dartmouth, USA
Sangwon Hyun   Myongji University, South Korea
Daisuke Mashima   Illinois at Singapore Pte. Ltd., Singapore
Sung Lee   VMware, USA
Sandra Céspedes   University of Chile, Chile
Mohammadreza Ashouri   University of Potsdam, Germany
Francesco Mercaldo   Università degli Studi del Molise, Italy
Carlos Rubio-Medrano   Arizona State University, USA
Wenjun Fan   University of Colorado, Colorado Springs, USA
Ihor Vasyltsov   Samsung Electronics, South Korea
Wei Yan   Clarkson University, USA
Ahyoung Lee   Kennesaw State University, USA
Daisuke Mashima   Advanced Digital Sciences Center, Singapore
Thomas Austin   San Jose State University, USA
Chang-Wu Chen   imToken, Taiwan
Hsiang-Jen Hong   University of Colorado, Colorado Springs, USA
Lei Xu   University of Texas Rio Grande Valley, USA
Hossein Sayadi   California State University, Long Beach, USA
Liudong Xing   University of Massachusetts Dartmouth, USA
Hyoungshick Kim   Sungkyunkwan University, South Korea
Attila Altay Yavuz   University of South Florida, USA
Donghyun (David) Kim   Georgia State University, USA
Jinoh Kim   Texas A&M University–Commerce, USA
Hongxin Hu   Clemson University, USA
Jorjeta Jetcheva   San Jose State University, USA
Tamzidul Hoque   University of Kansas, USA

Contents

Machine Learning for Security

Fake Malware Generation Using HMM and GAN

Harshit Trehan and Fabio Di Troia[✉]

San Jose State University, San Jose, CA 95192, USA
`fabio.ditroia@sjsu.edu`

Abstract. In the past decade, the number of malware attacks have grown considerably and, more importantly, evolved. Many researchers have successfully integrated *state-of-the-art* machine learning techniques to combat this ever present and rising threat to information security. However, the lack of enough data to appropriately train these machine learning models is one big challenge that is still present. Generative modelling has proven to be very efficient at generating image-like synthesized data that can match the actual data distribution. In this paper, we aim to generate malware samples as opcode sequences and attempt to differentiate them from the real ones with the goal to build fake malware data that can be used to effectively train the machine learning models. We use and compare different Generative Adversarial Networks (GAN) algorithms and Hidden Markov Models (HMM) to generate such fake samples obtaining promising results.

Keywords: Malware · Fake malware generation · GAN · HMM · Word embedding · Machine learning

1 Introduction

Malicious software, or malware in short, is a program that is specifically designed to harm computer systems by affecting devices, stealing or tampering with data, and even harming people. According to data collected by SonicWall, there were a total of 9.9 billion malware attacks worldwide in 2019 alone [29]. Thus, protection of computer systems from malware is an integral component of information security, and malware research plays an important role in securing computer systems.

To overcome these threats, machine learning techniques have been researched and applied in the malware detection domain. Their models are trained by extracting features such as opcode sequences, API calls, bytes vectors, and many other [1, 26, 32]. Although machine learning techniques have shown promising results, there are still some challenges to be taken in consideration, such as malware code obfuscation [22], the availability, or lack thereof, of large public datasets for the training phase [8], and adversarial machine learning [12] to deceive the machine learning models.

S.-Y. Chang et al. (Eds.): SVCC 2021, CCIS 1536, pp. 3–21, 2022.
https://doi.org/10.1007/978-3-030-96057-5_1

In this paper, we use mnemonic opcodes extracted from malware executable files belonging to five different malware families to generate realistic fake malware samples by implementing Generative Adversarial Networks (GAN) [9] and Hidden Markov Model (HMM). We use multiple machine learning classification techniques, namely, Support Vector Machines, k-Nearest Neighbor, Random Forest, and Naïve Bayes Classifier to differentiate between fake and real samples and compare the two techniques (HMM and GAN) based on their performance. The main goal of this project is to develop practical use cases for fake malware opcode sequences and serve as a *proof-of-concept* for using generative modelling to synthesize mnemonic opcode sequences. An embedding step is also introduced to convert the sequence of opcodes before being used to train the classifiers. While the majority of research in this field leaned towards creating fake malware images, this work introduces the creation of fake opcode sequences comparing HMM and different GAN variants.

The remainder of this paper is organized as follows. In Sect. 2, we go over previous and related work. Also, we give a brief summary of the techniques and concepts that we used in this paper. In Sect. 3, we explain our workflow and give a description of our malware generation pipeline. In Sect. 4, we go over the actual implementation and our experimental setup. In Sect. 5, we provide the results of our experiments. Finally, in Sect. 6 we discuss the results and the future directions for our project.

2   Background

In this Section, we discuss the background of malware classification and the use of generative modelling. We highlight the gap in the literature with respect to generated/synthetic malware opcode samples. We also give a brief introduction to Hidden Markov Models (HHMs) and Generative Adversarial Networks (GANs). Further reading about the machine learning techniques used to evaluate our results can be found at [5,7,27,28].

2.1   Background and Related Work

A recent trend in malware research is creating images from malware executable files and using them to perform malware detection and classification. This gives the opportunity to use image-analysis techniques, and allows for the use of powerful deep neural networks which perform exceptionally well with images [16,34].

In terms of generative modelling, many researchers used malware images to generate malware samples as that gives the advantage of boosting the dataset, and even performing data augmentation to real samples. For example, in [6] the authors adopted malware as images applying Variational Auto Encoder (VAE) and GANs to boost the malware dataset. They obtained a 2% and 6% increase in accuracy in case of, respectively, VAE and GAN. In another similar research [18], the authors used GAN and observed a 6% increase in accuracy using the benchmark ResNet-18 model trained on malware data.

Data augmentation or boosting using malware as images and generative modelling techniques is becoming increasingly popular. The drawback of this technique, though, is that converting malware files to images is computationally expensive. Moreover, training deep convolutional networks is also computationally expensive taking long time to train and test the models. Using GANs with images has similar overheads. An alternative solution is described in [11], where the authors propose a GAN based model, called "MalGAN", that is capable to bypass black-box malware detection systems whit almost 0% of detection rate. They used API features extracted from the malware samples as they are executed in a virtual environment. Despite of the impressive results, executing malware in a sandbox environment to extract the API features is again a not negligible overhead.

It is clear that there is a gap in the literature when it comes to generating malware samples using non-image features or representations of malware. Hence, we explore this gap by utilizing mnemonic opcodes extracted from malware files and generating mnemonic opcode sequences obtained by applying HMM and three different GAN architectures (see Sect. 2.3, 2.4, 2.5).

2.2 Hidden Markov Models

Hidden Markov Model (HMM) is a machine learning technique which is widely and effectively used for statistical analysis of time-series or sequential data. They have been successfully used in speech analysis and recognition [23], malware classification [2], and genes sequence analysis [17]. A Markov model is defined as a statistical model which has states and where the transition probabilities from one state to another are known. On the other hand, in an HMM the underlying states are not known to the observer. HMM, in fact, relies on the probability distribution of observing a set of observation symbols for each state [30].

We can use HMM to solve three Problems:

1. **Problem 1:** Given an observation sequence, $\mathcal{O}$, and a model λ, we can find $P(\mathcal{O}|\lambda)$. This means that we can compute a score for the sequence $\mathcal{O}$ w.r.t. λ [30].
2. **Problem 2:** Given a model λ and an observation sequence $\mathcal{O}$, we can determine the hidden states of the Hidden Markov Model. That is, we can uncover the Markov process underneath [30].
3. **Problem 3:** Given an observation sequence $\mathcal{O}$ and dimensions N and M, we can find the model λ of the given dimensions that best represents $\mathcal{O}$. This basically means that we are training the model to match the observation sequence [30].

The solution to these problems is implemented through the Baum-Welch algorithm [31]. In this paper, we solved all three of these problems, and more details are given in Sect. 4.2.

2.3   Generative Adversarial Networks

A Generative Adversarial Network (GAN) [9] model consists of two neural networks, the discriminator and the generator, which participate in a zero-sum game to achieve Nash equilibrium. The objective of the two networks is different from each other but the overall goal of the algorithm is to generate data samples that conform to a probability distribution p_g which is similar to the true data probability distribution p_{true}. The generator tries to fool the discriminator by forcing it to classify the generated samples as real, while the discriminator tries to correctly classify such samples. More information about the GAN working and architecture can be found in [3,9].

GAN Training. When actually training the model, the loss function used is Binary Crossentropy [20] which calculates the difference in the probability distribution of true samples, labelled 1, and false samples labelled 0. The weights of both models are updated independently of each other using two loss functions on the models parameterized by their weights.

More details about the GAN training algorithm can be found in [13].

GAN Limitations. Although GANs excel in learning complex data distributions, there exist major challenges in training GANs, such as mode collapse, vanishing gradient, internal covariate shift, failure mode, and more. To overcome these problems, several novel variants and architectures of GANs have been researched and implemented. The work in [15] and [19] provide a comprehensive analysis of the challenges in GAN training and the advantages and disadvantages of various GAN architectures.

2.4   Wasserstein GAN

Wasserstein GAN (WGAN) [4] was first proposed in 2017 by M. Arjovsky et al. as an improvement over the vanilla GAN. They first published a paper [3] highlighting the important theoretical implications of GAN training as proposed by Ian J. Goodfellow et al. [9], and outlined the mathematical reasoning and proofs for some of the issues surrounding GAN training.

WGAN Working. The main idea of the WGAN is that instead of optimizing the JS Divergence between two probability distributions, the use of a different distance metric as the loss function is proposed, that is, the Wasserstein distance or Earth-Mover distance. The Wasserstein distance is referred to as the Earth-Mover distance because it can be thought of as the minimum amount of energy cost required to transform the shape of a pile of dirt representing a probability distribution into the shape of another. The dirt is "transported" from one pile to another, and the cost is calculated as the mass moved times the distance. More details about this approach can be found in [4].

More details about the WGAN training algorithm can be found in [4].

WGAN Limitations. The main drawback of the WGAN algorithm is the way K-Lipschitz continuity is enforced [4]. Clipping the weights into a compact space $[-c, c]$ is not a very good way to enforce this constraint. It can lead to the model failing to learn more complex distributions and even saturating before reaching optimality. In fact, if the clipping parameter is large, it takes too much time for the weights to reach their limit and, thus, jeopardizing the training. On the other hand, if the clipping is small, we need to take in consideration the vanishing gradients problem.

2.5 WGAN with Gradient Penalty

Wasserstein GAN with Gradient Penalty (WGAN-GP) was first introduced in 2017 by Ishaan Gulrajani et al. [10]. The main objective of this architecture is to overcome the drawback of WGAN which is the way Lipschitz continuity is enforced.

To solve this, the authors in [10] propose an improved WGAN training method. They present Corollary 1 in [10] which claims that the optimal critic in WGAN has gradient norm equal to the value 1 and it is 1-Lipschitz continuous. Using this fact, a "penalty" is imposed on the critic if its gradient's norm deviates from the value 1. The training algorithm used in WGAN-GP is very similar to WGAN's algorithm minus the weight clipping part and the addition of the gradient penalty [10].

More details about the WGAN-GP training algorithm can be found in [10].

3   Methodology

In this Section, we detail our fake malware generation pipeline, feature extraction for fake sample evaluation, and the machine learning pipeline for our experiments.

3.1   Fake Malware Using HMM

The methodology adopted for generating fake malware samples using HMM is explained here:

1. Create observation sequence $\mathcal{O}$ of length $T = 30,000$ for each family.
2. Train 21 HMM models for each malware family with $T = 30,000, N = 2$ and $M \in \{20, 21, ..., 40\}$, where M is taken as top $M - 1$ most frequent opcodes and every opcode not present in top $M - 1$ was marked as "other" or M. Section 4 explains why we chose these values for M.
3. Score these 21 HMM models for each family by testing them against samples from the other four families and benign dataset.
4. Select the best value of M, say M', from these models for each family and train 10 HMM models by setting $N = M = M'$.
5. Score the 10 models for each family.

6. Select the two highest scoring models from Step 4 and use their γ matrix to find out the most likely state sequence of the HMM model. The most likely state sequence represents the fake samples.
7. Score and evaluate these fake samples as explained in Sect. 3.4.

3.2   Fake Malware Using GAN

We use three different GAN architectures to generate fake samples, that is, GAN, Wasserstein GAN (WGAN), and Wasserstein GAN with Gradient Penalty (WGAN-GP). The methodology adopted for generating fake malware samples using GANs is explained here:

1. Train GAN models for each family, and save generator models at an interval of 200 epochs for GAN and 500 for both WGAN and WGAN-GP.
2. Generate fake samples in batches of 32 using the saved generative models.
3. Evaluate them against real data samples by simply testing the integer vectors (Sect. 3.3) representing real samples and fake samples.
4. Repeat Step 4 five times and then average the results.
5. Select the best scoring model as the final generative model for each family, giving a total of five generator models per architecture.
6. The models selected in Step 6 are used to generate fake samples for each family and the samples are evaluated as explained in Sect. 3.4.
7. Repeat Steps 2–6 for WGAN and WGAN-GP architectures.

3.3   Feature Extraction

In this Section we explain our feature extraction process and the types of features used for evaluation. We extract three different features from the real and fake samples to train our machine learning models.

- **Normal integer vector conversion of opcodes:** We simply map the mnemonic opcodes to integers.
- **Word2Vec:** We treat the real samples as our corpus and create Word2Vec embedding of length 100 for each opcode. We use this embedding to create a vector for each data sample by simply summing up the embedding vector of each opcode in a given sample. Then, we normalize it by the length of the sample.
- **n-grams:** We create bigrams ($n = 2$) from the real dataset and find the top 20 bigrams based on the frequency. Then, a vector of length 20 is created for each data sample which contains the frequency count of these 20 bigrams. We treat these vectors as our bigram features.

3.4   Evaluation

We evaluated all the HMM models by creating the Receiver Operating Characteristic (ROC) curve for each model and calculating the Area Under the Curve

(AUC). For GAN, instead, we used a different approach because the most common application for GANs is in the image domain. However, we generate opcode sequences which can not be inspected visually. Hence, to evaluate our GAN models, we saved the generative model at every 200 epochs for GAN and 500 epochs for WGAN and WGAN-GP. From all the saved generative models we generated fake samples and classified them against real samples using Random Forest classifier. The model, identified by the epoch number, that gave the lowest classification results was chosen as the best generative model from that architecture, and then used for evaluation as explained in Sect. 4.

Accuracy, Precision and Recall. To score and evaluate the quality of the fake samples (HMM and GANs), we trained four machine learning models with each of the three features (Word2Vec, Bigram, integer vectors) and calculated the Accuracy, Precision, and Recall for each model. The process is explained here:

1. Randomly sample 100 real data samples and take 100 fake samples.
2. Extract features from real and fake samples as mentioned in Sect. 3.3.
3. Fit four different models, namely SVM, Random Forest, Naive Bayes classifier, and k-Nearest Neighbor on the training data using 5-fold cross validation.
4. Calculate the accuracy, precision and recall for each split done by 5-fold cross validation and use the average as the final result.

4 Implementation

In this Section, we give a detailed explanation of our dataset and the configuration of our HMM models, the different GAN approaches, and the machine learning techniques implemented for evaluation of our fake samples.

4.1 Dataset

Our dataset consists of five malware families and a benign dataset. Each malware family has over 1000 samples and the benign dataset has over 700 samples, both containing mnemonic opcode sequences. To build such dataset, we began with the Malicia dataset [21] which has over 50 malware families, and selected WinWebSec and Zbot families since these two has more than 1000 samples each. The rest of the three families were collected from VirusShare [24]. This dataset has over 120,000 malware executables and it is around 100 Gigabytes in size, from which we selected Renos, VBInject, and OnLineGames families.

We used `objdump` which is a command line program part of the GNU Binary Utilities library for Unix-like operating systems. This program is used to disassemble executables into Assembly code and, hence, to extract the mnemonic opcodes. Specifically, such code is processed via a Python script to remove all the unnecessary information such as registers, labels, and addresses, to obtain sequences containing only the opcodes found in the code. A summary of our dataset along with each malware family's type is given in Table 1.

4.2   HMM Implementation

The HMM algorithm was implemented following the algorithm given in [30]. We wrote the code in C++, with the addition of an external Python script to preprocess our data and create the observation sequence $\mathcal{O}$ of length $T = 30,000$. We concatenated the mnemonic opcodes from different samples of a family until we reached a length of $30,000$. This was done for all five families in our dataset.

Table 1. Dataset summary

Malware family	Type	Samples
Benign	Benign samples	706
OnLineGames	Password stealer	1513
Renos	Trojan Downloader	1568
VBInject	Worm	2694
WinWebSec	Rogue	4360
Zbot	Password stealer	2136

The number of unique opcodes for each family was very high and setting M to such large values makes training of HMM models computationally infeasible. Thus, we experimented with selecting the top n most frequent opcodes from the observation sequence, where $n \in \{20, 21, ..., 40\}$. The value n is represented as the parameter M in HMM, and its optimal value for each family served as the dimensions of our HMM model in the next set of experiments ($N = M = M'$).

Afterwards, we solved Problem 2 of HMM to find the most likely state sequence which will act as our fake malware samples generated using HMM. For each family, our model dimensions were $N \times M$, where $N = M = M'$ and M' was the best value of M for each individual family.

We trained ten different HMM models, each with 5000 random restarts for each malware family. All ten of these models were scored the same way as explained above, using 500 true samples and 500 false samples. Out of these ten models, we selected the two best ones with the highest AUC value. The γ matrix from these two models was used to find the most likely hidden state sequence. Each model gives us a sequence of 30,000 length. Finally, we divided this sequence into 50 "fake" samples of length 600 each. This gives us a total of 100 fake samples per family.

4.3   GAN Implementation

We implemented all three GAN architectures in Python using TensorFlow and Keras with TensorFlow backend. For GAN, we used Adam optimizer with the following parameters:

$$Adam(lr = 0.0003, \ \beta_1 = 0.5, \ \beta_2 = 0.99)$$

These parameters gave the best results and, thus, they were chosen. The loss function used was Binary Crossentropy as it is equivalent to the loss function for GAN. The models were trained for 10000 epochs.

For GANs, the use of Batch Normalization [14] layer is recommended as the training is done using minibatches of data. The variance in the input data implicitly caused by minibatches slows down training and requires the use of very small learning rates, otherwise the gradients and weights of layers may change drastically from minibatch to minibatch. For the discriminator we have one input layer, two fully connected hidden layers, and an output layer with just one neuron. The activation function for the output layer is Sigmoid since we are using Binary Crossentropy loss, and Sigmoid gives a value between $[0, 1]$ which is interpreted as the score for a sample or the probability. The activation function for the hidden layers is LeakyReLU. LeakyReLU is recommended over ReLU because ReLU outputs 0 for all negative inputs which causes vanishing gradients problem. LeakyReLU has the hyperparameter α which is used to scale negative outputs. We used $\alpha = 0.2$ for our experiments. LeakyReLU activation function is:

$$f(x) = \begin{cases} \alpha x & x \leq 0 \\ x & x > 0 \end{cases} \tag{1}$$

The generator has one input layer, three fully connected hidden layers with a batch normalization layer after every hidden layer, and finally an output layer with 600 neurons, which is the length of the opcode sequence we want to generate. The activation function for hidden layers is, again, LeakyReLU, and for the output layer we used TanH. We scale all of our inputs between $[-1, 1]$, and TanH also gives an output between that range, which is what we expect from the generator. We experimented with different layers for both networks, including Convolutional 1D layers, and fully connected Dense layers had the best performance.

GAN Stabilizing Techniques. We further utilized stabilizing techniques to improve GAN training. All the techniques are discussed in [25] which was published in 2016 by some of the co-authors of the original 2014 paper on GANs [9]. The techniques were Minibatch Discrimination, Label Smoothing, and Label Switching.

4.4   WGAN Implementation

For WGAN, we used RMSProp optimizer. RMSProp is recommended by the paper authors in [4] because the training was more stable for RMSProp as compared to Adam which is momentum based. The learning rate chosen is also a small value:

$$RMSProp(lr = 0.00001)$$

The architecture of our WGAN is the same for the critic and the generator, except the input and output layers. We trained each WGAN model for 100,000 epochs using minibatches of data.

The actual models are compiled and trained separately for the critic and generator. For the generator, we have the same activation function for hidden layers (LeakyReLU) and output layer (TanH). For the critic, however, we used no activation function or used linear activation in the output layer. This approach allows the loss function to be computed easily when implementing the WGAN algorithm given in [4]. These layers and networks gave the best result, hence, we chose these as our final networks.

4.5   Wasserstein Distance

The loss function or the Wasserstein distance between real and fake samples can be written as follows:

$$\text{Critic loss} = \text{critic's avg. real samples score - critic's avg. fake samples score}$$
$$\text{Generator loss} = \text{- critic's avg. fake samples score}$$

This interpretation is correct because we want the critic network to learn the K-Lipschitz function that will calculate the Wasserstein distance. We are only concerned with the output of the function and not actually knowing the function. Assuming the network has learnt the correct function, we can interpret the Wasserstein distance as the loss given above.

Since neural networks use stochastic gradient descent they seek to minimize the loss values. For the generator, minimizing the loss value will mean that the critic will be encouraged to score the fake samples higher. For example, a score of 5 on fake samples will mean -5 loss for the generator and a score of 10 will mean -10 loss. For the critic, in order to minimize the loss, the score for real samples will be encouraged to be small. This will maximize the distance between the generated and fake samples and at the same time minimize both losses. This is implemented simply by using no activation function in the output layer for the critic and using -1 label for fake samples and $+1$ for real samples.

4.6   WGAN with Gradient Penalty Implementation

For WGAN with Gradient Penalty, we used Adam optimizer. Unlike WGAN, momentum based optimizers seem to work well for WGAN-GP. The parameters for the optimizer were:

$$Adam(lr = 0.0001, \; \beta_1 = 0.5, \; \beta_2 = 0.9)$$

We trained each WGAN-GP model using minibatches for 100,000 epochs. We decided to use Convolutional 1D layers for the models because using fully connected Dense layers had worse performance as compared to Conv1D layers. In the critic network, we used three hidden Conv1D layers with 64, 128, and 256 filters and filter size 3. In the generator network, we also used three Conv1D layers with 64, 32, and 16 filters, and filter size 3. The activation functions for the hidden Conv1D layers is again LeakyReLU.

The output layer of the generator is a fully connected Dense layer with 600 neurons, and the activation function is again TanH. Similar to WGAN, the output layer of the critic network has no activation function because we still need to calculate the Wasserstein loss/distance. The authors in [10] advised against the use of Batch Normalization in the critic network. They suggested that, if required, Layer Normalization could be used. We experimented with Layer Normalization but the performance degraded, hence, we decided not to implement it. For the generator, we still used Batch Normalization layer.

We used $\lambda = 10$, that is, the penalty coefficient, and the parameter $n\_critic = 7$, that is, the number of critic iterations per generator iteration. Additionally, after every 500 epochs, we trained the critic for 100 iterations and, then, updated the generator. This allows for exact Wasserstein distance calculation instead of an approximation and, therefore, the generator receives the correct gradient updates to converge properly.

5 Results and Discussion

In this Section, we discuss and present the results of our experiments.

5.1 HMM Results

The first set of experiments were conducted to determine the optimal value of M for each family. Then next set of experiments were conducted to train the best HMM models which were used to generate fake malware samples. The summary of the results and the best value of M chosen for each family can be found in [33].

For HMM models to generate fake samples by solving Problem 2, we fixed the dimensions as $N = M$, where M is the best value for each family.

Our next experiments consisted of training ten different HMM models with dimensions as mentioned above and choose the two best models out of ten. We chose the two highest scoring models and calculated their most likely hidden state sequence using the γ matrix from the models. After breaking the two γ matrices of 30,000 length each into 100 samples of length 600 each, we tested these fake samples against real samples as explained in Sects. 3.3, 3.4. Due to low accuracy, precision, and recall scores, the model was not able to differentiate between real and fake samples. Results from each of the four algorithms are given in the following Section.

HMM Classification Results. We first performed hyperparameter tuning for the four machine learning algorithms and fixed the best parameters for the rest of the experiments.

1. **SVM:** Grid search on the values of C, kernel, and degree with ranges: $C \in \{1, 2, \ldots, 10\}$, $kernel \in \{rbf, poly, linear\}$, and $degree \in \{2, 3, 4, 5\}$.

We found that polynomial kernels were overfitting the data, hence, the final parameters for SVM were $C = 5$ and kernel $= rbf$.

2. **Naïve Bayes:** No hyperparameter tuning required for Naïve Bayes classifier.
3. **Random Forest:** Grid search on the number of decision trees to use, and maximum depth of trees, with ranges: number of trees $\in \{10, 20, \ldots, 80\}$, max depth of trees $\in \{2, 3, \ldots, 10\}$. We found that using 50 decision trees with max depth of 5 performed best without overfitting the real malware samples.
4. **k-NN:** Grid search on the number of neighbors to consider (k) with range: $k \in \{4, 5, \ldots, 20\}$. The value $k = 8$ worked well, and the distance metric chosen was Euclidean.

We used 5-fold cross validation and the scores given are the average scores from 5-fold cross validation. By using Word2Vec features, SVM, Random Forest, and k-NN classifiers, we were able to differentiate between real and fake samples efficiently. Especially SVM with accuracy, precision, and recall equal to 1.00 for all the families, except Zbot with 0.97, 0.99, and 0.95, respectively. However, Naïve Bayes classifier had low recall rates for Zbot (0.73) and OnLineGames (0.76). We attribute this result to the ineffectiveness of the classifier rather than the quality of fake samples.

When Bigram features were applied, all four classifiers were able to differentiate between real and fake samples very effectively with accuracy, precision, and recall in between 0.97 and 1.00, with the only exception of OnLineGame with accuracy and precision rates equal to 0.96 and 0.93 when Naïve Bayes classifier and k-NN were used.

Finally, by using integer vectors, the metrics rates were less consistent, varying between 0.59 and 1.00, with particularly poor results when Naïve Bayes and k-NN classifiers were used. We attribute these low scores to integer vectors being a weaker feature representation for the data.

5.2   GAN Results

We experimented with the stabilizing techniques mentioned in Sect. 4.3. Although the training stabilized across all five families using these techniques, the results improved for Zbot, Renos, and VBInject but got worse for WinWebSec and OnLineGames. This is a common phenomenon when training GANs. The loss values for the discriminator and generator do not necessarily indicate or correspond to the model's performance or quality of the generated samples. Fake samples were generated using the best chosen models in batches of 32 since that was the batch size during training. Generating samples in same batch sizes as the training size, generally, gives better results.

We used the same hyperparameters as discussed in Sect. 5.1, and tested the fake samples using all three features mentioned above.

Using Word2Vec and Bigram features, the scores for all four families dipped a little as compared to the HMM results. SVM and Random Forest reached accuracy, precision, and recall above 0.90 for these two features, except for OnLineGames with 0.88 precision with Random Forest. Low precision rate means high false positive rate which was the most desirable result for us. Naïve Bayes had low overall scores for Word2Vec and Bigram features on account of it being a weaker classifier. Interestingly, k-NN obtained the lowest overall scores for these two features. This can be attributed to the way k-NN algorithm works and that the generated data distribution is slightly closer to the real data distribution as compared to HMM fake samples.

For integer vectors, we found that all four classifiers were not able to effectively differentiate between real and fake samples. As seen with the previous experiments, integer vectors are a weaker feature representation but the difference in results between HMM integer vector classification and GAN integer vector classification does suggest that the GAN models were able to perform better than HMM. For k-NN and Renos, the precision and recall are 0%, which means that the model was not able to distinguish between fake and real at all based on just the integer vectors.

5.3   WGAN Results

Unlike GAN, the loss values when training WGAN gave reliable information about the model's progress and convergence. Hence, for WGAN and WGAN-GP, we first discussed the loss curves and convergence and, then, gave the classification results for the four machine learning techniques.

Convergence and Loss Values. The loss value for the critic and the generator converged very fast in the first few epochs and, then, stayed the same for the remaining epochs. We tried a lot of different hyperparameters, such as changing the value of "n_critic", different clipping value, and different learning rates. Even changing the networks entirely and using Convolutional 1D instead of fully connected Dense layers did not help. The value of loss did not change after the first few epochs. This shows that clipping the weights is a major drawback in WGAN (Sect. 2.4) as it saturates the model, and the weights do not update after a point. Any change in weight is nullified by the clipping step. Interestingly, all four families converged to the same loss value for the critic and generator. The clipping step stops the training since the weights can not change beyond the clipping range and do not respond to the gradient updates that are back propagated through the network.

WGAN Classification Results. The best generative model from WGANs was chosen independently for each family. We used the same hyperparameters as discussed in Sect. 5.1, and tested the fake samples using all three features in batches of 32.

Using Word2Vec and Bigram features, SVM and Random Forest were able to effectively differentiate between real and fake samples generated by WGAN, with accuracy, precision, and recall ratios between 0.96 and 1.00 for all the families. Interestingly, even Naïve Bayes and k-NN performed well, even though we found from the previous results that they were the two weaker classifiers. This means that the WGAN fake samples were of inferior quality compared to HMM and GAN.

Using integer vectors, the results for SVM and Random Forest were high (in between 0.85 and 1.00), but not as effective as the ones obtained with Word2Vec and Bigram features. Again, integer vectors proved to be a weak feature representation that makes classification hard. For k-NN and Naïive Bayes with integer vectors, we obtained extremely low recall rates for some families, such as 0.25 for VBInject, 0.37 for OnLineGames, and 0.53 for Renos. However, these low recall rates were accompanied by high precision rates of almost 1.00 across all the families.

5.4   Wasserstein GAN with Gradient Penalty

As with WGAN, the critic's loss value helps monitor the model's performance for WGAN-GP. The WGAN-GP paper [10] mentions that the the critic's loss should start at a large number and then converge towards zero. The generator's loss is not very insightful and can fluctuate. Thus, first we discuss the loss curves and then give the classification results.

Convergence and Loss Curves. The loss curves for all five families showed a similar shape, with the start value for the critic that started at around -28 and then slowly converged to around -4. This is the expected behavior and means that our model was training properly.

The critic loss curves for the other four families also showed similar shapes but with slightly different values of convergence. Training the models for more epochs, around 200,000–300,000, would be ideal for full convergence.

The loss curve for the generator was not very informative about the model's performance and training, as the loss values kept oscillating.

WGAN-GP Classification Results. The best generative model from WGAN-GPs was chosen independently for each family. We used the same hyperparameters as discussed in Sect. 5.1, and tested the fake samples using all three features in batches of 32.

Using Word2Vec and Bigram features, all four machine learning techniques were not able to give very good classification results. Compared to WGAN and GAN, the metrics rates were much lower (in between 0.77 and 1.00). This means that the quality of the fake samples generated by WGAN-GP generative models is better as compared to WGAN and GAN. The most surprising result is the dip in Random Forest's classification. Random Forest is one of the better classifiers out of the four classifiers that we used. For Zbot, Renos, and VBInject the overall accuracy for Random Forest was around 0.70. For WinWebSec and OnLineGames, the accuracy was also low at 0.82 and 0.81 for Word2Vec, respectively, and even lower for Bigram features at 0.81 and 0.74. This is a promising result since we saw that classifying real and fake samples using these two features was very effective, getting high accuracy and precision scores previously.

Using integer vector features the scores for SVM, Naïve Bayes, and k-NN classifiers were very low (in between 0.00 and 0.90). These three models were not able to distinguish between real and fake samples just based on the integer representation. This was confirmed by accuracy scores in range of 0.50 and 0.60, and even lower for Naïve Bayes at less than 0.50 for WinWebSec, Zbot, Renos, and VBInject families. Random Forest did a better job as compared to the other three techniques but the accuracy was still around 0.70 for WinWebSec, Zbot, Renos, and around 0.60 for OnLineGames and VBInject. This again showed that the quality of fake samples generated by WGAN-GP generative model was much better than the other GAN architectures and HMM.

5.5 Comparison of the Results

The complete results for the WGAN-GP experiments can be found in [33]. In Fig. 1, we compare the four different approaches by computing the average accuracy per malware family over the three different embeddings. We can see that the sequences generated by HMM and WGAN techniques are the ones more easily detected, that is, their generated fake malware is not confused with the real malware data. GAN obtains better results but they are not far from the previous ones. WGAN-GP, on the other hand, is the approach that clearly shows its potential in confusing the classifiers. In fact, the average accuracy obtained with the four classifiers is consistently poor. This shows the difficulty in detecting the fake WGAN-GP data from the real one.

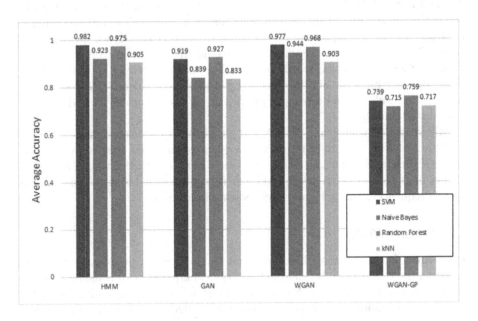

Fig. 1. Comparison of the results

6   Conclusion and Future Work

In this paper, we aimed at utilizing different generative modelling techniques to generate fake malware mnemonic opcode sequences. We utilized four different techniques, that is, Hidden Markov Models (HMMs), Generative Adversarial Networks (GANs), Wasserstein Generative Adversarial Networks (WGANs), and Wasserstein Generative Adversarial Networks with Gradient Penalty (WGAN-GP).

We used three different feature extraction techniques to generate the malware opcode sequences, namely, Word2Vec, Bigram, and integer vectors. Classification results showed that Word2Vec and Bigram features gave a better representation of the malware data since for all four generative models the classification results were superior. Integer vectors, on the other hand, do not capture the true distribution of the real malware samples.

Fake samples generated by HMM were quite effectively distinguishable by SVM, Random Forest, and k-NN classifiers. Especially by using Word2Vec and Bigram features, these three classifiers obtained accuracy above 0.90 for all five of the tested families. Naïve Bayes classifier, instead, had much lower scores with any of the three feature extraction techniques.

Using generative models from GAN, we saw a slight improvement in the results with the fake malware being confused in larger number with the legitimate ones. For WGAN, the results were instead not promising. In fact, the classifiers were able to identify the fake malware samples with scores close to the ones obtained in the HMM experiments. This was attributed to the weight clipping step in the WGAN algorithm, that inhibits the critic network's ability to properly learn the real data's representation. However, for WGAN-GP we got the best results. We saw that the classification outcome was now relatively poor, even when the more informative Word2Vec and Bigram features were applied. In fact, for all four classifiers, we obtained accuracy in between 0.70 and 0.82. For integer vectors the results were even more promising, as the accuracy score dipped to around 0.50 and 0.60.

We concluded that using WGAN-GP algorithm is the best approach to successfully generate fake malware opcode sequences such that they appear closer to the real data distribution. This serves as a *proof of concept* that GAN algorithms, in particular WGAN-GP, can be successfully applied to generate malware opcode sequences, and not only in generating image data.

6.1   Future Work

There are a lot of different directions that this paper can be expanded in. For example, the dataset can be enlarged and the experiments can consider a larger number of malware families. Furthermore, instead of training individual GAN models for each family, a multi-class generative model can be considered. Another possible application is to use trained generative models to boost or augment the datasets for families that have a limited number of data samples. Other GAN variants could also be considered and compared, such as EBGAN and LSGAN.

Finally, experiments with LSTM-GAN can be conducted since stateful networks can provide interesting results.

References

1. Ahmadi, M., Ulyanov, D., Semenov, S., Trofimov, M., Giacinto, G.: Novel feature extraction, selection and fusion for effective malware family classification. In: Proceedings of the Sixth ACM Conference on Data and Application Security and Privacy, pp. 183–194 (2016)
2. Annachhatre, C., Austin, T.H., Stamp, M.: Hidden Markov models for malware classification. J. Comput. Virol. Hacking Tech. **11**(2), 59–73 (2014). https://doi.org/10.1007/s11416-014-0215-x
3. Arjovsky, M., Bottou, L.: Towards principled methods for training generative adversarial networks (2017)
4. Arjovsky, M., Chintala, S., Bottou, L.: Wasserstein Gan (2017)
5. Biau, G., Scornet, E.: A random forest guided tour. TEST Official J. Spanish Soc. Stat. Oper. Res., 197–227 (2016). https://doi.org/10.1007/s11749-016-0481-7
6. Burks, R., Islam, K.A., Lu, Y., Li, J.: Data augmentation with generative models for improved malware detection: a comparative study*. In: 2019 IEEE 10th Annual Ubiquitous Computing, Electronics Mobile Communication Conference (UEMCON), pp. 0660–0665 (2019). https://doi.org/10.1109/UEMCON47517.2019.8993085
7. Cortes, C., Vapnik, V.: Support-vector networks. Mach. Learn. **20**, 273–297 (1995)
8. Gibert, D., Mateu, C., Planes, J.: The rise of machine learning for detection and classification of malware: research developments, trends and challenges. J. Network Comput. Appl. **153**, 102526 (2020)
9. Goodfellow, I.J., et al.: Generative adversarial networks (2014)
10. Gulrajani, I., Ahmed, F., Arjovsky, M., Dumoulin, V., Courville, A.: Improved training of wasserstein GANs (2017)
11. Hu, W., Tan, Y.: Generating adversarial malware examples for black-box attacks based on Gan (2017)
12. Huang, L., Joseph, A.D., Nelson, B., Rubinstein, B.I., Tygar, J.D.: Adversarial machine learning. In: Proceedings of the 4th ACM Workshop on Security and Artificial Intelligence, pp. 43–58 (2011)
13. Hui, J.: Gan - what is generative adversarial networks GAN? December 2019. https://jonathan-hui.medium.com/gan-whats-generative-adversarial-networks-and-its-application-f39ed278ef09
14. Ioffe, S., Szegedy, C.: Batch normalization: accelerating deep network training by reducing internal covariate shift (2015)
15. Jabbar, A., Li, X., Omar, B.: A survey on generative adversarial networks: Variants, applications, and training. ArXiv abs/2006.05132 (2020)
16. Jain, M.: Image-based malware classification with convolutional neural networks and extreme learning machines, December 2019. https://scholarworks.sjsu.edu/etd_projects/900/
17. Krogh, A.: An introduction to hidden Markov models for biological sequences. In: Salzberg, S., Searls, D., Kasif, S. (eds.) Computational Methods in Molecular Biology, pp. 45–63. Elsevier, London (1998)
18. Lu, Y., Li, J.: Generative adversarial network for improving deep learning based malware classification. In: 2019 Winter Simulation Conference (WSC), pp. 584–593 (2019). https://doi.org/10.1109/WSC40007.2019.9004932

19. Pavan Kumar, M.R., Jayagopal, P.: Generative adversarial networks: a survey on applications and challenges. Int. J. Multimedia Inf. Retrieval **10**(1), 1–24 (2020). https://doi.org/10.1007/s13735-020-00196-w

20. Mannor, S., Peleg, D., Rubinstein, R.: The cross entropy method for classification. In: Proceedings of the 22nd International Conference on Machine Learning, ICML 2005, pp. 561–568. Association for Computing Machinery (2005). https://doi.org/10.1145/1102351.1102422

21. Nappa, A., Rafique, M.Z., Caballero, J.: The MALICIA dataset: identification and analysis of drive-by download operations. Int. J. Inf. Secur. **14**(1), 15–33 (2015)

22. O'Kane, P., Sezer, S., McLaughlin, K.: Obfuscation: the hidden malware. IEEE Secur. Priv. **9**(5), 41–47 (2011). https://doi.org/10.1109/MSP.2011.98

23. Rabiner, L.: A tutorial on hidden Markov models and selected applications in speech recognition. Proc. IEEE **77**(2), 257–286 (1989). https://doi.org/10.1109/5.18626

24. Roberts, J.M.: VirusShare.com - Because Sharing is Caring (2011). http://www.virusshare.com

25. Salimans, T., Goodfellow, I., Zaremba, W., Cheung, V., Radford, A., Chen, X.: Improved techniques for training GANs (2016)

26. Santos, I., Penya, Y.K., Devesa, J., Bringas, P.G.: N-grams-based file signatures for malware detection. ICEIS **2**(9), 317–320 (2009)

27. Sawla, S.: Introduction to Naïive Bayes for classification (2018). https://medium.com/@srishtisawla/introduction-to-naive-bayes-for-classification-baefefb43a2d

28. Scikit-learn: K Neighbors Classifier. https://scikit-learn.org/stable/modules/generated/sklearn.neighbors.KNeighborsClassifier.html. Accessed 09 May 2021

29. SonicWall: Sonicwall 2020 Cyber Threat Report (2020). https://www.sonicwall.com/news/2020-sonicwall-cyber-threat-report

30. Stamp, M.: A revealing introduction to hidden Markov models. Science, 1–20 (2004)

31. Stamp, M.: Introduction to Machine Learning with Applications in Information Security, 1st edn. Chapman & Hall/CRC (2017)

32. Sun, Z., et al.: An opcode sequences analysis method for unknown malware detection. In: ICGDA 2019, pp. 15–19. Association for Computing Machinery (2019)

33. Trehan, H.: Fake malware opcodes generation using HMM and different GAN algorithms. Master's thesis, San Jose State University (2021). https://scholarworks.sjsu.edu/etd_projects/1001/

34. Yajamanam, S., Selvin, V.R.S., Di Troia, F., Stamp, M.: Deep learning versus gist descriptors for image-based malware classification. In: ICISSP, pp. 553–561 (2018)

Security Threats in Cloud Rooted from Machine Learning-Based Resource Provisioning Systems

Hosein Mohammadi Makrani[1]([✉]) [iD], Hossein Sayadi[2] [iD], Najmeh Nazari[1] [iD], and Houman Homayoun[1] [iD]

[1] University of California, Davis, USA
{hmakrani,nnazaribavarsad,hhomayoun}@ucdavis.edu
[2] California State University, Long Beach, USA
hossein.sayadi@csulb.edu

Abstract. Resources provisioning on the cloud is problematic due to heterogeneous resources and diverse applications. The complexity of such tasks can be reduced with the aid of Machine Learning. Researchers have found, however, that machine learning poses new threats such as adversarial attacks. Based on our investigation, we found that adversarial ML can target resource provisioning systems (RPS) to perform distributed attacks. Our work proposes a fake trace generator (FTG), which can be wrapped around an adversary kernel to avoid detection by the RPS and to enable the adversary to get co-located with the victim's virtual machine.

Keywords: Machine-learning · Cloud · Resource-provisioning

1 Introduction

Due to the rise of social media, Internet-of-Things (IoT), and multimedia, the volume of data has increased continuously, resulting in an overwhelming amount of data known as big data. In order to efficiently process such massive data, scale-out architecture has gained interest as a promising solution that is designed to provide a massively scalable computer architecture. Recent improvements in the networking, storage, energy-efficiency and infrastructure management have made cloud (the best example of scale-out architecture) a preferable approach to respond to the new computing challenges.

A resource provisioning system provides various services including resource efficiency [11], security, fault tolerance, and monitoring to achieve the performance goals while maximizing the utilization of available resources [10] in the cloud. The latest recourse provisioning systems, which they were successful to significantly improve the utilization, used machine learning techniques to overcome the challenge of diversity of applications and heterogeneity of resources in the cloud.

RPSs routinely schedule multiple applications from multiple users on the same physical hosts to increase efficiency, in a way that applications have minimum impact on each other's performance. Moreover, a recent work proposed

S.-Y. Chang et al. (Eds.): SVCC 2021, CCIS 1536, pp. 22–32, 2022.
https://doi.org/10.1007/978-3-030-96057-5_2

to exploit information used by resource provisioning systems for scheduling purposes, for detecting an adversary VM by its micro-architectural trace and behavior. In this way, they are actually adding another line of defense, this time in the scheduling phase, against attackers.

On the other hand, the interference on shared resources from multi-tenancy can lead to security and privacy vulnerabilities. Interference may leak important information ranging from a service's placement to confidential data, like private keys [3]. This has prompted significant work on distributed side-channel [9] and distributed denial of service attacks (DDoS) [6], data leakage exploitations [22], and attacks that pinpoint target VMs in a cloud system [19]. However, none of the above attacks targeted the resource provisioning system by itself to use it as a new point of vulnerability and a platform for their attacks. Most of those attacks leverage the lack of strictly enforced resource isolation between co-scheduled instances and the naming conventions cloud frameworks use for machines to extract confidential information from victim applications, such as encryption keys.

In this work, we show how utilizing machine learning in resource provisioning systems can become a blind spot and weakness to be exploited by adversaries for planning an attack. Despite the machine learning systems being deployed in numerous applications and shown robustness against random noises [18], the exposed vulnerabilities have shown that the outcome of ML models can be modified or controlled by adding specially crafted perturbations to the input data, often referred to as Adversarial samples. A plethora of works on adversarial attacks exists, focusing specifically on computer vision applications, where the number of features is often large. Recently, a few works on crafting adversarial traces are as well proposed [12].

We argue that the adversarial samples in ML can be leveraged to impose security risk and manipulate today's ML-based RPSs by reverse engineering the ML models from the performance and utilization data these systems generate. We show an example (DDoS attack) to how an adversary can bypass the instance initialization phase of RPS and get co-located by victims with high probability. We also will show how it is possible to disguise the malicious behavior of the adversary's VM and still remain on the same host with the victim and avoid the migration. To create such a fake trace generator, we use the concept presented in [12] for the adversarial sample generation in machine learning. By reverse engineer the resource provisioning system, we can create an adversarial sample for the adversary's application trace. We run FTG as a separate thread inside the adversarial VM and by expecting the transferability of such an attack, we improve the effectiveness of distributed attacks.

2   Security Threats

We show that ML solutions hides security vulnerabilities, since it enables an adversary to extract information about an application's type and characteristics. An adversarial VM has the goal of disguising as Friendly VM to determine the nature and characteristics of any applications co-scheduled on the same physical host, and negatively impact their behavior.

2.1 Threat Model

Our work focuses on IaaS providers that offer public clouds to mutually untrusting customers where multiple VMs can be co-located on the same server. VMs do not have any control over where they are placed, nor do they have any information about other VMs on the same physical host. As a result, at this point, we assume that the resource provisioning system will be neutral with respect to detection of adversarial virtual machines, which means that it won't assist such attacks or employ additional resource isolation techniques to prevent them.

Adversarial VM: Adversary virtual machines are designed to steal information or negatively impact the performance of the victims by getting co-located with them and evading detection mechanisms embedded in resource provisioning systems.

Friendly VM: One or more applications are run on this virtual machine, which is scheduled on a physical host. No techniques for preventing detection, such as obfuscation of memory patterns, are used.

2.2 Distributed Attack

A distributed attack [1] goal is to retrieve secret information, or decrease the performance of computing nodes on a distributed system, where each computing node processes a part of the overall data. The examples of retrieved information may be a set of encryption keys that can be used to compromise the functionality of the whole distributed system. A distributed attack may also be used to retrieve information about the cloud infrastructure such as FPGA cartography and fingerprinting. In the following, we present some characteristics and provide more details of such attacks.

Definition 1. We can define the distributed attack over a set M_{vic} of virtualized instances running in a distributed system S, as a tuple $DSCA = (S, M_{vic}, D, M_{mal}, A, CP, EP)$ where: S is a distributed system; M_{vic} are the VMs that are targeted by the attack; D is the distributed dataset to be compromised (partially or totally); M_{mal} are malicious VMs, co-located with the victim VMs; A is a set of local attack techniques (such as side channel [16], denial of service, or resource freeing attack); CP is a protocol to coordinate the attacker VMs in M_{mal}; EP is a protocol to exfiltrate data.

We consider $D = d_1, ..., d_n$ a dataset to be processed by the distributed system $S = s_1, ..., s_n$ implemented on a set of VMs $M_{vic} = m_{vic1}, ..., m_{vicn}$ on a virtualized platform. Each component s_i of S processes data d_i locally and runs in its own VM m_{vici}. To perform the distributed attack, the adversary sets up a number of malicious VMs(at least equal to the number of M_{vic}) $M_{mal} = m_{mal1}, ..., m_{maln}$, co-located with the victim instance M_{vic}. The adversary also masters a set $A = a_1, ..., a_m$ of local attack techniques, i.e., Flush+Reload.

The objective of a distributed attack is to first attack each component of the system s_i running on m_{vici} through m_{mali} running local attack technique a_j to retrieve d_i. The synchronization between attack instances and a central server may be performed using a coordination protocol CP. A protocol EP may be used to control attacking instances remotely, and to send collected information to a remote server to exfiltrate sensitive data. In the following, we briefly explain three well-known local attack on a distributed system:

Side Channel Attack. By sharing physical resources like processor caches, or by using virtualization mechanisms, side-channels may occur due to lack of enforced isolation. The side channel is a hidden information channel that is different from the main channel (e.g., network), in that the protection mechanisms around the data might not be adequate to prevent security violations. The purpose of a side channel attack is to exploit a side channel for obtaining critical information. Side channel attacks can be classified according to the type of exploited channel. The two most popular types of SCA are timing attacks and cache-based attacks, where the cache memory of the processor is often exploited by adversaries.

Denial of Service Attack. The overloading of server resources caused by a denial of service attack degrades the performance of the victim service. These attacks can be classified as external or internal (or host-based) in cloud settings specifically. IaaS cloud multi-tenancy allows internal DoS attacks to launch adversarial programs on the same host as the victim and impact its performance.

Resource Freeing Attack. In addition, resource-freeing attacks (RFAs) hurt the victim's performance as well as forcing them to surrender resources to the adversary [17]. Despite their effectiveness, RFAs require significant compute and network resources, and are subject to defenses, like live VM migration.

2.3   Attack's Setting: VM Co-location

An adversarial VM is rarely interested in a random service running on a public cloud. They need to pinpoint where the target resides in a practical manner to be able perform DoS, RFA, or SC attacks. This requires a launch strategy and a mechanism for co-residency detection. The attack is practical if the target is located with high accuracy, in reasonable time and with modest resource costs. We show that by black box attack to the RPS's model and eventually generating adversarial sample, we can force the RPS to put the Adversarial VM on the desired host. Once a target VM is located, the adversary can launch RFA, or DoS attack.

2.4   Locating Physical Hosts Running Victim Instances

In order to accomplish co-residency with the victim instance, an attacker needs to launch several VMs. This is impractical and not feasible. As side-channel and

RFA attacks are local attacks, it is essential that the malicious VMs reside on the same physical host as the victim VMs. Finding the physical hosts running virtual machines on which the victims are running is therefore the first and most important step. It is important to consider factors such as datacenter region, instance type, and time interval when aiming for co-residency. Among IaaS clouds, these variables may vary. The application type is, however, considered an important factor in placement [21]. Let $P(m_{mali})$ be the probability of instance m_{mali} to be co-resident with instance victim m_{vici}. The value of P will be raised by increasing the number of launched attack instances. To make sure that both attacker and victim VMs achieve coresident placement, the adversary can perform co-residency detection techniques such as network probing [7]. The attacker can also use data mining techniques to detect the type and characteristics of a running application in the victim VM by analyzing interferences introduced in the different resources to increase the accuracy of co-residency detection.

2.5   Avoidance of Detection and Migration

In virtualized environments, there are several techniques for detecting attacks. A side-channel attack, for example, would require very fine-grained information in order to be detected [15]; this information can primarily be provided by Hardware Performance Counters (HPCs) [13]. Modern microprocessors contain a set of special-purpose registers called HPCs that capture hardware events such as last-level cache (LLC) load misses, branch instructions, branch misses, and executed instructions while executing an application. Events of this type are primarily used for analyzing program behavior and are accessible to everyone in the user space. Detection of abnormalities in computer systems is also based on these events. We distinguish two different methods of detection: (1) signature based [14] and (2) threshold-based [2]. The signature-based approach generates a signature of the attack based on information received from HPCs and compares the behavior of the system with the generated signature to identify if any malicious activity has been detected. On the other hand, threshold-based approaches utilize the HPCs trace to flag anomaly resource utilization that goes beyond a pre-specified threshold.

3   ML Based Resource Provisioning System

Figure 1 shows how a normal ML based RPS works. First, they monitor the application and extracts its micro architectural information. Then based on the current behavior and server configuration, they generate a performance model for the application. By leveraging an optimization techniques and available cost model, they determine the suitable configuration and host for the application. In this study we use PARIS [20], a ML based performance model proposed at Berkeley as a cost aware resource provisioning system.

PARIS uses Random Forest for predicting performance from the application fingerprint to find the best VM type configuration. To generate the fingerprint of application, PARIS extracts 20 resource utilization counters spanning the following broad categories and calls it fingerprint: CPU utilization, Network utilization, Disk utilization, Memory utilization, and System-level features. On the other hand, CPU count, core count, core frequency, cache size, RAM per core, memory bandwidth, storage space, disk speed, and network bandwidth of the server are the representation of the configuration provisioned by PARIS.

We denote the micro architectural fingerprint and system level information of an application as $Finger\_print$ vector. In Eq. (1), f_i denotes each architectural feature.

$$Finger\_print = \{f_1, f_2, ..., f_{20}\} \tag{1}$$

configuration parameters of the server platform referred to configuration inputs is as follow:

$$Configuration = \{c_1, c_2, ..., c_9\} \tag{2}$$

where $Configuration$ is the configuration vector and c_i is the value of the ith configuration parameter (number of sockets, number of cores, core frequency, cache size, memory capacity, memory frequency, number of memory channel, storage capacity, storage speed, network bandwidth).

The RPS is responsible to provision $Configuration$ based on $Finger\_print$:

$$Configuration = f(Finger\_print) \tag{3}$$

Note that $f(Finger\_print)$ is just a data model, which means there is no direct analytical equation to formulate it.

3.1 Reverse Engineering the Model

As mentioned, RPS can be considered as a blackbox (worst-case scenario). In such cases, we perform a reverse engineering to mimic the functionality of the RPS. Thus, as a first step to craft adversarial malware, we perform reverse engineering similar to that proposed in [8].

In order to reverse engineer, we first create a training dataset that comprises of all types of applications. Nearly 11,000 applications are used in the reverse engineering process. The Original RPS is fed with all the applications and the responses are recorded. These responses are utilized to train different ML classifiers in order to mimic the functionality of the original RPS. Further, it is tested by comparing the outputs from original RPS response and the reverse engineered RPS's response. Reverse engineering is non-trivial as the adversaries generated on a closely functional model will be highly effective compared to a weakly generated adversary. To ensure the reverse engineering is performed in an efficient way, we train multiple ML classifiers and choose the classifier that yields high accuracy.

Table 1. Detailed information of local cluster

Server (Xeon)	Freq. (GHz)	Socket	Core	Cache (MB)	Mem. (GB)	Storage	Server type	Count
E5-4669 V4	2.2	4	22	55	96	SSD PCIe	HPC	2
E5-4667 V4	2.2	4	18	45	64	SSD SATA	HPC	2
E5-4650 V4	2.2	4	14	35	32	SSD SATA	HPC	2
E5-2690 V4	2.6	2	14	35	512	SSD/HDD	Memory opt.	4
E5-2650 V4	2.2	2	12	30	256	SSD/HDD	Memory opt.	4
E5-2667 V4	3.2	2	8	25	32	SSD PCIe	I/O opt.	4
E5-2643 V4	3.4	1	6	20	32	SSD PCIe	I/O opt.	4
E5-2660 V2	2.2	2	10	25	16	HDD	General purp.	6
E5-2650 V2	2.6	2	8	20	16	HDD	General purp.	6
E5-1630 V4	3.7	1	4	10	8	HDD	Power opt.	2
E5-1680 V4	3.4	1	8	20	12	HDD	Power opt.	2
E3-1270 V6	3.8	1	4	8	8	HDD	Power opt.	2

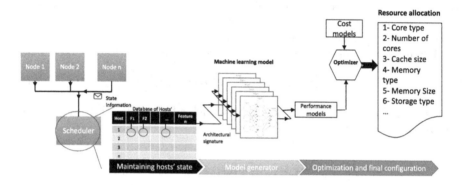

Fig. 1. ML based resource provisioning system

We perform the data collection in a controlled environment, where all applications are known. We use a 40-machine cluster (presented in Table 1), and schedule a total of 120 workloads, including batch analytics in Hadoop and Spark and latency-critical services, such as webservers, Memcached and Cassandra. For each application type, there are several different workloads with respect to algorithms, framework versions, datasets, and input load patterns. The training set is selected to provide sufficient coverage of the space of resource characteristics. The selected workloads cover the majority of the resource usage space.

We submit all of these applications to RPS. In the beginning, the RPS profiles the application and extracts the fingerprint. Then, the RPS uses the Random Forest model to determine an appropriate server configuration. We collect all the fingerprints and their correspondent configurations generated by the RPS to shape our dataset.

3.2   Adversarial Sample Generator

Once the reverse engineered RPS is built, it is non-trivial to determine the level of perturbations that need to be injected into application's micro architectural patterns in order to get the desired host configuration. The micro architectural patterns are perturbed by applying a gradient loss based approach, similar to the Fast-Gradient Sign Method (FGSM), which is widely used in image processing. The low complexity and low computation overheads are the benefits of such an approach. To train our neural network, we use reverse engineered ML RPS i.e., neural network with θ as the hyper parameters, x being the input to the model, and y is the output for a given input x, and $L(\theta, x, y)$ be the cost function used to craft adversarial perturbations. Using the gradient of the cost function of the neural network, the perturbation required to change the output to the target configuration is calculated. The adversarial perturbation generated based on the gradient loss, similar to the FGSM [5] is given by:

$$x^{adv} = x + \epsilon \, sign(\nabla_x L(\theta, x, y)$$

where ϵ is a scaling constant ranging between 0.0 to 1.0 is set to be very small such that the variation in $x(\delta x)$ is undetectable. In case of FGSM the input x is perturbed along each dimension in the direction of gradient by a perturbation magnitude of ϵ. Considering a small ϵ leads to well-disguised adversarial samples that successfully fool the ML model. In contrast to the images where the number of features are large, the number of features i.e., micro architectural metrics are limited, thus the perturbations need to be crafted carefully and also be made sure it can be generated during runtime by the applications. For instance, a negative value cannot be generated by an application. Hence, we provided lower bound on the adversary values. [4] presented how to craft the adversarial application so as to generate the perturbations during runtime.

3.3   Case Study

To evaluate our proposed approach, we implemented 8 distributed attacks as follow: SC1: Prime+Probe, SC2: Flush+Reload, SC3: Flush+Flush, SC4: Evict + Time, DoS1: increasing latency by saturating the network, DoS2: decreasing throughput by saturating storage, RFA1: freeing memory resource, and RFA2: freeing CPU resource. We perform these attacks on 20 unseen victim applications from different domains (SPEC, Hadoop, Spark, Memcache, and Cassandra). Based on our evaluation, the success rate of being co-located with victims, evading the detection and migration, and getting the desired outcome from attack depends on many factors such as victim's type, the period of monitoring phase, and amount of perturbation.

We now perform the DoS attack on utilization. If it causes the resource saturation, DoS will be detected and the victim will be migrated to a new machine. Our cluster supports live migration. Figure 2 compares the tail latency and CPU utilization with adversarial VM to that of a naive DoS that saturates the CPU through a CPU-intensive task. It shows the adversarial attack does not saturate the resource and does not cause the migration while still can put pressure on the victim.

Table 2. Effectiveness of distributed attacks based on application type

	SPEC	Hadoop	Spark	Memchahced	Cassandra
SC1	***	*	**	*	**
SC2	***	*	*	*	**
SC3	***	*	**	**	*
SC4	***	**	**	*	**
DoS1	*	*	***	***	**
DoS2	*	***	*	*	***
RFA1	*	**	***	***	**
RFA2	***	**	***	***	*

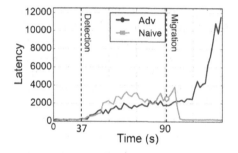 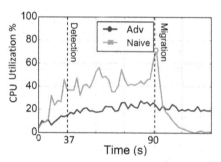

Fig. 2. Latency and utilization with adversarial sample and a naive DoS attack that saturates memory resources

Table 2 shows the impact of victims' type on the success rate of each type of attack. The interesting observation is that there is a meaningful relationship between the application's type and the nature of the attack by itself. For instance, we observe that side-channel attacks are more successful when the cache hit rate of the victim is low. Similarly, we observed that RFA is more successful when the resource utilization of the victim is high. One reason is that in such case FTG can generate a better fake trace to convince the RPS to stay at the current host. In a case that the difference between the behavior of the adversary kernel and the victim is high, the FTG has to generate more perturbation and this may lead to a migration decision by RPS.

4 Conclusions

The proposed adversarial attack on RPS comprises of three phases. Firstly, we perform reverse engineering to build a ML RPS that mimics the functionality of the original RPS. Further, with the aid of adversarial sample generator, the micro architectural pattern required to obtain the target server configuration is determined. Lastly, this crafted adversarial micro architectural pattern generator is spawned as separate thread, leading to overall pattern close to the victim VM's pattern, and eventually causes to be co-located with it. This means without saturating the resources or act as an abnormal application, by generating only small noise in applications behavior, we can force RPS to co-locate the adversary VM with victim and also fool the RPS to change the resources required for the targeted VM and impacts on its behavior. The goal of this study is to encourage public cloud providers to implement more stringent isolation solutions for their platforms and system engineers develop robust RPSs to deliver predictability and security at high utilization levels.

References

1. Bazm, M.-M., Lacoste, M., Südholt, M., Menaud, J.-M.: Isolation in cloud computing infrastructures: new security challenges. Ann. Telecommun., 197–209 (2019). https://doi.org/10.1007/s12243-019-00703-z
2. Chiappetta, M., Savas, E., Yilmaz, C.: Real time detection of cache-based side-channel attacks using hardware performance counters. Appl. Soft Comput. **49**, 1162–1174 (2016)
3. Delimitrou, C., Kozyrakis, C.: Quasar: resource-efficient and QoS-aware cluster management. In: ACM SIGARCH Computer Architecture News, vol. 42, pp. 127–144. ACM (2014)
4. Dinakarrao, S.M.P., et al.: Adversarial attack on microarchitectural events based malware detectors. In: DAC (2019)
5. Goodfellow, I.J., et al.: Explaining and harnessing adversarial examples. arXiv preprint arXiv:1412.6572 (2014)
6. Gupta, S., Kumar, P.: VM profile based optimized network attack pattern detection scheme for DDOS attacks in cloud. In: Thampi, S.M., Atrey, P.K., Fan, C.-I., Perez, G.M. (eds.) SSCC 2013. CCIS, vol. 377, pp. 255–261. Springer, Heidelberg (2013). https://doi.org/10.1007/978-3-642-40576-1_25
7. İnci, M.S., Gulmezoglu, B., Eisenbarth, T., Sunar, B.: Co-location detection on the cloud. In: Standaert, F.-X., Oswald, E. (eds.) COSADE 2016. LNCS, vol. 9689, pp. 19–34. Springer, Cham (2016). https://doi.org/10.1007/978-3-319-43283-0_2
8. Khasawneh, K.N., et al.: RHMD: evasion-resilient hardware malware detectors. In: MICRO (2017)
9. Liu, F., Ren, L., Bai, H.: Mitigating cross-VM side channel attack on multiple tenants cloud platform. JCP **9**(4), 1005–1013 (2014)
10. Makrani, H.M., et al.: Adaptive performance modeling of data-intensive workloads for resource provisioning in virtualized environment. ACM Trans. Model. Perform. Eval. Comput. Syst. (TOMPECS) **5**(4), 1–24 (2021)
11. Makrani, H.M., Sayadi, H., Motwani, D., Wang, H., Rafatirad, S., Homayoun, H.: Energy-aware and machine learning-based resource provisioning of in-memory analytics on cloud. In: Proceedings of the ACM Symposium on Cloud Computing, pp. 517–517 (2018)
12. Makrani, H.M., et al.: Cloak & co-locate: adversarial railroading of resource sharing-based attacks on the cloud. In: 2021 IEEE International Symposium on Secure and Private Execution Environment Design (SEED). IEEE (2021)

13. Mucci, P.J., Browne, S., Deane, C., Ho, G.: PAPI: a portable interface to hardware performance counters. In: Proceedings of the Department of Defense HPCMP Users Group Conference, vol. 710 (1999)
14. Payer, M.: HexPADS: a platform to detect "Stealth" attacks. In: Caballero, J., Bodden, E., Athanasopoulos, E. (eds.) ESSoS 2016. LNCS, vol. 9639, pp. 138–154. Springer, Cham (2016). https://doi.org/10.1007/978-3-319-30806-7_9
15. Sayadi, H., et al.: Towards accurate run-time hardware-assisted stealthy malware detection: a lightweight, yet effective time series CNN-based approach. Cryptography **5**(4), 28 (2021)
16. Sayadi, H., et al.: Recent advancements in microarchitectural security: review of machine learning countermeasures. In: 2020 IEEE 63rd International Midwest Symposium on Circuits and Systems (MWSCAS), pp. 949–952. IEEE (2020)
17. Varadarajan, V., Kooburat, T., Farley, B., Ristenpart, T., Swift, M.M.: Resource-freeing attacks: improve your cloud performance (at your neighbor's expense). In: Proceedings of the 2012 ACM Conference on Computer and Communications Security, pp. 281–292. ACM (2012)
18. Wang, H., Sayadi, H., Sasan, A., Rafatirad, S., Mohsenin, T., Homayoun, H.: Comprehensive evaluation of machine learning countermeasures for detecting microarchitectural side-channel attacks. In: Proceedings of the 2020 on Great Lakes Symposium on VLSI, pp. 181–186 (2020)
19. Xu, Z., Wang, H., Wu, Z.: A measurement study on co-residence threat inside the cloud. In: 24th USENIX Security Symposium (USENIX Security 15), pp. 929–944 (2015)
20. Yadwadkar, N., et al.: Selecting the best VM across multiple public clouds: a data-driven performance modeling approach. In: ACM SoCC (2017)
21. Zhang, W., et al.: A comprehensive study of co-residence threat in multi-tenant public PaaS clouds. In: Lam, K.-Y., Chi, C.-H., Qing, S. (eds.) ICICS 2016. LNCS, vol. 9977, pp. 361–375. Springer, Cham (2016). https://doi.org/10.1007/978-3-319-50011-9_28
22. Zhang, Y., Juels, A., Reiter, M.K., Ristenpart, T.: Cross-VM side channels and their use to extract private keys. In: Proceedings of the 2012 ACM Conference on Computer and Communications Security, pp. 305–316. ACM (2012)

Differential Privacy in Privacy-Preserving Big Data and Learning: Challenge and Opportunity

Honglu Jiang[✉] , Yifeng Gao , S. M. Sarwar , Luis GarzaPerez ,
and Mahmudul Robin

The University of Texas Rio Grande Valley, Edinburg, TX 78504, USA
{honglu.jiang,yifeng.gao,sm.sarwar01,luis.garzaperez,
mahmudul.robin01}@utrgv.edu

Abstract. Differential privacy (DP) has become the de facto standard of privacy preservation due to its strong protection and sound mathematical foundation, which is widely adopted in different applications such as big data analysis, graph data process, machine learning, deep learning, and federated learning. Although DP has become an active and influential area, it is not the best remedy for all privacy problems in different scenarios. Moreover, there are also some misunderstanding, misuse, and great challenges of DP in specific applications. In this paper, we point out a series of limits and open challenges of corresponding research areas. Besides, we offer potentially new insights and avenues on combining differential privacy with other effective dimension reduction techniques and secure multiparty computing to clearly define various privacy models.

Keywords: Differential privacy · Deep learning · Big data

1 Introduction

Organizations, companies and governments collect data from a variety of sources, including social networking, transactions, smart Internet of Things devices, industrial equipment, electronics commercial activities, and more, which can be used to dig out valuable information hidden behind the massive data for modern life. The extensive collection and further processing of personal information in the context of big data analytics and machine learning-based artificial intelligence results in serious privacy concerns. For example, in March 2018, Facebook-Cambridge Analytica was reported to use the personal data of millions of people's Facebook profiles harvested without their consents for political advertising purposes in the 2016 US presidential election, which was a great political scandal and caused an uproar in the world. Despite the benefits of analytics, it cannot be accepted that big data comes at a cost for privacy. Therefore, the present study shifts the discussion from "big data versus privacy" to "big data with privacy", adopting the privacy and data protection principles as an essential value [5]. Privacy-preserving data publishing (PPDP) and various artificial

© Springer Nature Switzerland AG 2022
S.-Y. Chang et al. (Eds.): SVCC 2021, CCIS 1536, pp. 33–44, 2022.
https://doi.org/10.1007/978-3-030-96057-5_3

intelligence-empowered learning/computing have gained significant attentions in both academia and industry. It is, thus, of utmost importance to craft the right balance between making use of big data technologies and protecting individuals' privacy and personal data [5].

Intuitively, one can make use of the simple naive identity removal to protect data privacy, but in practice, it does not always work. For instances, AOL released an anonymized partial three-month search history to the public in 2006. Although personally identifiable information was carefully processed, some identities were accurately reidentified. For example, *The New York Times* immediately located the following individual: the person with number 4417749 was a 62-year-old widowed woman who suffered from some diseases and has three dogs. Such real-world privacy leakage problems and attack instances clearly demonstrate the importance of data privacy preservation.

The problem of data privacy protection was first put forward by Dalenius in the late 1970s [6]—Dalenius pointed out that the purpose of protecting private information in a database is to prevent any user (including legitimate users and potential attackers) from obtaining accurate information about arbitrary individuals. Following that, many privacy preservation models with strong operability including k-anonymity, l-diversity [20], t-closeness [18] were proposed. However, each model generally provides protection against only a specific type of attacks and cannot defend against newly developed ones. A fundamental cause of this deficiency lies in that the security of a privacy preservation model is highly related to the background knowledge of an attacker. Nevertheless, it is almost impossible to define the complete set of possible background knowledge an attacker may have.

Dwork originally proposed the concept of *differential privacy* (DP) to protect against the privacy disclosure of statistical databases in 2006 [4]. Under differential privacy, query results of a dataset are insensitive to the change of a single record. That is, whether a single record exists in the dataset has little effect on the output distribution of the analytical results. As a result, an attacker cannot obtain accurate individual information by observing the results since the risk of privacy disclosure generated by adding or deleting a single record is kept within an acceptable range. Unlike anonymization model, DP makes the assumption that an attacker has the maximum background knowledge, which rests on a sound mathematical foundation with a formal definition and rigorous proof.

It is worth noting that differential privacy is a definition or standard for quantifying privacy risks rather than a single tool, which is widely used in statistical estimations, data publishing, data mining, and machine learning. It is a new and promising privacy framework and has become a popular research topic in both academia and industry, which can be potentially implemented in various application scenarios. However, DP is a strict privacy standard, the data utility is likely to be poor while providing a meaningful privacy guarantee. The goal of this paper is to summarize and analyze the state-of-the-art research and investigations in the field of differential privacy and its applications in privacy-preserving data publishing, machine learning, deep learning, and federated learning, to point out a series of limits and open challenges of corresponding research areas, so as to provide some approachable strategies for researchers and engineers to implement DP in real world applications. In our paper, we place more focus on practical applications of differential privacy rather than detailed theoretical analysis of differentially private algorithms.

The rest of this paper is organized as follows. We present the background knowledge of differential privacy in Sect. 2. Section 3 introduces differentially private data publishing problem and presents some challenges on this problem. In Sect. 4, we summarize existing research on the application of differential privacy to deep learning and federated learning. Section 5 concludes the paper with some future research discussion and open problems on differential privacy applications.

2   Preliminary of Differential Privacy

Differential privacy can be achieved by injecting a controlled level of statistical noise into a query result to hide the consequence of adding or removing an arbitrary individual from a dataset. That is, when querying two almost identical datasets (differing by only one record), the results are differentially privatized in that an attacker cannot glean any new knowledge about an individual with a high degree of probability, i.e., whether or not a given individual is present in the dataset cannot be guessed.

2.1   Definition of Differential Privacy

Let f be a query function to be evaluated on a dataset D. Algorithm A runs on the dataset D and sends back $A(D)$. $A(D)$ could be $f(D)$ with a controlled amount of random noise added. The goal of differential privacy is to make $A(D)$ as much close to $f(D)$ as possible, thus ensuring data utility (enabling the user to learn the target value as accurately as possible), while preserving the privacy of the individuals with the added random noise. The main procedure can be seen in Fig. 1.

Definition 1 *(Neighboring Datasets). Two datasets D and D' are considered to be neighboring ones if $d(D, D') = 1$, where $d(D, D')$ is the number of records D and D' differ.*

Definition 2 *(Differential Privacy [8]). A randomized algorithm A is (ϵ, δ)-differentially private if for any two datasets D and D' with $d(D, D') = 1$, and for all sets S of possible outputs, we have*

$$Pr[A(D) \in S] \leq e^{\epsilon} Pr[A(D') \in S] + \delta,$$

where ϵ and δ are non-negative real numbers.

When $\delta = 0$, the algorithm becomes ϵ-differentially private. We say a mechanism gives δ-approximate differential privacy when $\delta \neq 0$. The ϵ is often a small

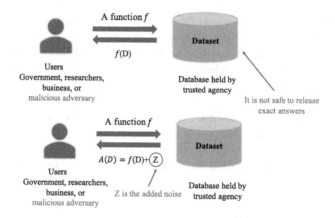

Fig. 1. The framework of differential privacy

positive real number called *privacy budget*, which is used to control the probability of the algorithm A getting almost the same outputs from two neighboring datasets. It reflects the level of privacy preservation that algorithm A can provide. For example, if we set $\epsilon = \ln 2$, the result S is at most twice as likely to be generated by dataset D as by any of D's neighbor D'.

The smaller the ϵ, the higher the level of privacy preservation. A smaller ϵ provides greater privacy preservation at the cost of lower data accuracy with more additional noise. When $\epsilon = 0$, the level of privacy preservation reaches the maximum, i.e., "perfect" protection. In this case, the algorithm outputs two results with indistinguishable distributions but the corresponding results do not reflect any useful information about the dataset. Therefore, the setting of ϵ should consider the trade-off between privacy requirements and data utility. In practical applications, ϵ usually takes very small values such as 0.01, 0.1, or $\ln 2$, $\ln 3$.

2.2   Noise Mechanism of Differential Privacy

Sensitivity is the key parameter to determine the magnitude of the added noise, that is, the largest change to the query result caused by adding/deleting any record in the dataset. Accordingly, global sensitivity, local sensitivity, smoothing upper bound, and smoothing sensitivity are defined under the differential privacy model. Because of the limitation of space, we will specifically introduce them here.

(1) Laplace Mechanism
The Laplace distribution (centered at μ) with scale b is the distribution with probability density function

$$h(z) = \frac{1}{2b} \exp(-\frac{|z - \mu|}{b}).$$

Let $Lap(b)$ denote the Laplace distribution (centered at 0) with scale b.

Definition 3 *(Laplace Mechanism [8]). For dataset D and function $f : D \to R^d$ with global sensitivity GS_f, the Laplace mechanism $A(D) = f(D) + Z$ is ϵ-differentially private, where $Z \sim Lap(GS_f/\epsilon)$.*

The Laplace mechanism is suitable for the protection of numerical results. Taking an example Laplace mechanism for the counting function, since the global sensitivity of counting is 1, that is $GS_f = 1$, if we choose $\epsilon = 0.1$, the Laplace mechanism outputs $3 + Lap(10)$.

(2) Exponential Mechanism
The Laplace mechanism is appropriate only for preserving the privacy of numerical results. Nevertheless, in many practical implementations, query results are entity objects. McSherry *et al.* put forward the exponential mechanism [21] for the situations where the "best" needs to be selected. Let the output domain of a query function be *Range*, and each value $r \in Range$ be an entity object. In the exponential mechanism, the function $q(D, r)$, which is called the *utility function* of the output value r, is employed to evaluate the quality of r.

Definition 4 *(Exponential Mechanism [21]). Given a random algorithm A with the input dataset D and the output entity object $r \in Range$, let $q(D, r)$ be the utility function and Δq be the global sensitivity of function $q(D, r)$. If algorithm A selects and outputs r from Range at a probability proportional to $\exp(\frac{\epsilon q(D,r)}{2\Delta q})$, then A is ϵ-differentially private.*

2.3   Local Differential Privacy

Traditional centralized differential privacy provides privacy protection based on a premise that there is a trusted third-party data collector who does not steal or disclose user's sensitive information, while local differential privacy [7] does not assume the existence of any trusted third-party data collector. Instead, it transfers the process of data privacy protection to each user, making each user independently deal with and protect personal sensitive information.

Definition 5 *(Local Differential Privacy [7]). Given n users, with each corresponding to a record. A privacy algorithm M with definition domains $Dom(M)$ and $Ran(M)$ satisfies the ϵ-local differential privacy if M obtains the same output result t^* ($t^* \subseteq Ran(M)$) on any two records t and t' ($t, t' \in Dom(M)$):*

$$Pr[M(t) = t^*] \leq e^\epsilon \times Pr[M(t') = t^*]$$

One can see from this definition that local differential privacy provides privacy by controlling the similarity between the output results of any two records, while each user processes its individual data independently, that is, the privacy preserving process is transferred to a single user from the data collector, such that a trusted third party is no longer needed and privacy attacks brought from the data collection of untrusted third-party is thus avoided. The framework of local differential privacy can be seen in Fig. 2.

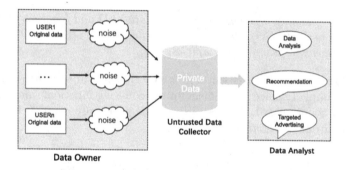

Fig. 2. The framework of local differential privacy

3 Differentially Private Data Publishing

3.1 Differential Privacy in Tabular Data Publishing

The goal of differentially private data publishing is to output aggregate/synthetic information to public without disclosing any individual's information. Generally, there are two settings in the data publishing scenario, interactive and non-interactive. In the first setting, users make queries request to the data curator, who answers the query with a noisy result. The fixed privacy budget will be exhausted as the number of queries increases. In the non-interactive setting, the data curator publishes statistical information related to the dataset that satisfies differential privacy. When the queries are submitted, the corresponding query result is directly returned from the published synthetic dataset.

The challenge of interactive setting is that the number of queries is limited while the privacy budget ϵ is easily exhausted. That is, a higher accuracy result for one query with less noise results and a larger ϵ usually results in a smaller number of queries.

High sensitivity presents a big challenge on the data publishing in the non-interactive setting, while high sensitivity means large magnitude of noise and low data utility especially for big data and complex data, which we will detailed introduce in Sect. 3.3. Another problem is that the published synthetic dataset can only be used for particular purposes or targeted a fixed query function.

3.2 Differential Privacy in Graph Data Publishing

With the widespread application of social networks, the increasing volumes of user-generated data have become a rich source which can be published to third parties for data analysis and recommendation system. Generally, social networking data can be modeled as graph $G(V, E)$, where V is a set of nodes and E is a set of relational activities between nodes. Analyzing graph data such as analysis of social network data has great potential social benefits and help generate insights into the laws of data change and trend characteristics. Most popular tasks of social network analysis include degree distribution, subgraph counting (triangle counting,

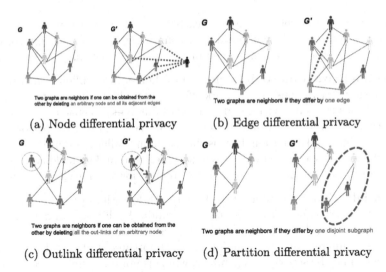

(a) Node differential privacy   (b) Edge differential privacy

(c) Outlink differential privacy   (d) Partition differential privacy

Fig. 3. Differential privacy definitions in graph data

k-star counting, k-triangle counting,etc.) and edge weight analysis. In reality, various types of privacy attacks such as de-anonymization attacks [11,12,14,15,23], inference attacks [10,16] on social networks have raised the stakes for privacy protection while a large amount of personal user data have been exposed.

However, the privacy issue of graph is more complicated starting from how to model and formalize the notion of "privacy" in graph network. Differential privacy originates from tabular data, while the key to extending differential privacy to social networks is to determine the neighboring input entries, that is, how to define "adjacent graphs". Figure 3 shows existing definitions of DP in graph data, namely, node differential privacy, edge differential privacy, outlink differential privacy, partition differential privacy; detailed information can be referred to [13].

3.3   Challenges on Differentially Private Data Publishing

In this subsection, we present a few challenges and open problems on differentially private data publishing especially for big data, complex network, dynamic and continuous data publishing.

As what it reads, big data deal with massive amounts of data at a great speed passing, which exhibit various characteristics that cover challenges like gathering, analysis, storage and privacy preservation. Of the many characteristics of big data, *5V* characterizes big data's nature the best, namely Volume, Velocity, Variety, Veracity and Value.

Differential Privacy on Complex and High Volume Network Structure. Network structures such as social networks and traffic networks are often complex. Since query sensitivities are usually high, much noise has to be added to query results to achieve differential privacy. Nevertheless, the noise may significantly affect the output data utility, resulting in useless data. Moreover, it may be hard to effectively compute sensitivities, either global or smooth, precise or approximate, as the computational complexity may be too high (or even NP-hard) to be practical for many complex graph network analysis queries.

Differential Privacy on High Dimensional Data. Most differentially private data publishing techniques cannot work effectively for high dimensional data. On one hand, since the sensitivities and entropy of different dimensions vary, evenly distributing the total privacy budget to each dimension degrades the performance. Moreover, "Curse of Dimensionality" is the common challenge in big data perturbation which means a dataset contains high dimensions and large domains resulting in a pretty low "Signal-to-Noise" and extremely low data utility even useless.

Differential Privacy on Correlated Data. Differential privacy offers a neat privacy guarantee while it is a strict privacy standard, while assumes all the data are independent, while the correlation or dependence may undermine the privacy guarantees of differential privacy mechanisms. Unfortunately, the real-world gathered data can not be strictly independent, which is not only tuple (record) correlated but also attribute information correlated. For example, the salary information in strongly correlated with education level and occupation in a dataset.

Differential Privacy on High-Velocity Data. Velocity in big data refers to the crucial characteristic of capturing data dynamically. In practical applications, the data are dynamically updated such as recommendation system, trajectory data to capture the evolutionary behaviors of various users. Differential privacy on continuous flow of data faces critical challenges of great noise accumulation and privacy budget allocation for each time sequence.

4 Differentially Private Machine Learning

4.1 Differential Privacy in Deep Learning

The privacy protection provided by DP also could benefit the existing deep learning model. Generally, the noise can be added into the gradient, input, and embedding. Adadi et al. [1] introduce the first DP preserved optimization algorithm named DPSGD. The DP is achieved by adding Gaussian noise in every SGD optimization step. Arachchige et al. [2] introduce a model named LATENT. The framework achieves the protection by transferring the real-vale low dimensional representation into a discrete vector. Lecuyer et al. [17] proposed a model named PixelDP. The framework achieves the goal by adding Gaussian noise in the hidden layers of a CNN model. Different from these works, Phan et al. [22] proposed a method that directly manipulates the inputs. The model induces different levels of noise for each pixel of an image based on a relevant score [3].

4.2 Differential Privacy in Federated Learning

The research field of Federated Learning focuses on learning a model where data is stored in a distributed system. As pointed out by Wei et al. [25], attackers can retrieval the data information through the gradient, a DP preserved learning model could protect such information leakage in the Federated Learning setting. Wei et al. [24] integrated DP algorithm into the Secure Multiparty Computation(SMC) framework. DP is used to encrypt the response for each query in the SMC. Geyer et al. [9] introduce a DP algorithm focusing on removing the data source info. In addition to using the same SGD algorithm framework as DPSGD, the algorithm also will randomly ignore a portion of the data to protect data privacy.

4.3   Challenges on Differentially Private Machine Learning

Model Dependency. Other than the gradient-based approach, most deep-learning based DP algorithms introduced in this paper are highly related to the deep learning model. For example, LATENT and PixelDP are designed only for CNN. A DP approach that does not rely on the data and model could be promising research in the DP research field.

Accuracy Loss of Federated Learning Due to Added Noise. In federated learning model, differential privacy-based approaches add noise to the uploaded parameters which will degrade the model accuracy inevitably and further affect the convergence of the global aggregation. Moreover, there are few results about practical frameworks integrating differential privacy and other cryptography-based methods, which hinders the industrial development of federated learning.

5   Future Directions and Conclusions

Differential privacy is a strong standard of privacy protection with a solid mathematical definition which can be applied in various application scenarios, however differential privacy is not a panacea for all privacy problems and the research on differential privacy is still in its infancy stage. There are still some misunderstandings, inappropriate applications and flawed implementations in differential privacy. In this section, we propose a few future research problems and open problems that worthy of more attention.

5.1   Combination of Differential Privacy and Other Technologies

As we mentioned about the privacy preservation of high dimensional data, it is feasible and promising to combine effective dimensionality reduction techniques with differential privacy to address this issue. Specifically, it is possible to try both linear and non-linear transformation such as compressive sensing and manifold learning which maps a high-dimensional space to a low-dimensional representation.

With the great high privacy concern on Federated learning, IoT network and other distributed environment, the combination of local differential privacy, multiparty computations and sampling and anonymization will be a future topic which needs open-ended exploration. Secure multiparty computation is a type of

cryptography-based which could be concerning and infeasible on computationally constrained devices, while anonymization model has its own shortcomings
about the assumption on background knowledge. However, the combination of
these techniques can boost the performance of differential privacy. Specifically,
differential privacy with a sampling processing can greatly amplify the privacy
preservation level [19], based on which we can adapt the idea of anonymization
to participants of DP processing. For example, in the scenario of federated learning, we can randomly pick up the clients and parts of differentially private local
updates to form a shuffle model. Moreover, inter-discipline techniques between
local differential privacy and secure multiparty computation involve the secure
computation, privacy preservation and dataset partition, which need to tackle
with the high communication cost and low data utility.

5.2 Variation of Differential Privacy and Personalized Privacy

Differential privacy provides strong and strict privacy guarantee at the cost of
low data utility while it may be too strong and not necessary in some practical
applications. To achieve a better tradeoff between privacy and preservation, various relax and extensions of differential privacy need to be proposed and in fact
many of these definition have been proposed such as *crowd-blending privacy, individual differential privacy*, and *probabilistic indistinguishability*. However, most
of these are still in the stage of theoretical definition or be specific scenarios.
The great challenge is that how to widely apply to these extensions to practical
applications.

On the other hand, conventional private data privacy preservation mechanisms aim to retain as much data utility as possible while ensuring sufficient
privacy protection on sensitive data while such schemes implicitly assume that
all data users have the same data access privilege levels. Actually, data users
often have different levels of access to the same data, personalized requirements
of privacy preservation level or data utility. It is a big challenge to achieve personalized privacy and multi-level data utility while the uniform framework itself
is a hard problem.

5.3 Misunderstandings of Differential Privacy vs More Than Privacy

As we mentioned in differentially private data publishing, the data utility of
outputs are likely to be very poor or with large privacy budget, that is lower privacy preservation level, which we cannot sure how much privacy it can provides.
Moreover, when differential privacy is applied to federated learning, it is used on
local updates of parameters while traditional differential privacy is designed for
record data contributed by different individuals on the basis of assumption that
the data are independent. However, in federated/distributed learning, all local
data are from the same client which have little possibility to be independent.

In contrast, differential privacy can do more while there exists misconceptions
and misuse of differential privacy. Besides providing privacy preservation through

hiding individual information in the aggregate information, from the opposite perspective of its definition, differential privacy can ensure that the probability of outcomes unchanged when modifying any individual record in the training data, and the application of this property needs to be explored. Secondly, differential privacy can also protect against the malicious attacks in learning techniques such as poisonous attacks in federated learning which can help improve the accuracy of training model. Thirdly, specific differentially private methods can be combined with reward mechanisms in distributed learning to provide privacy preservation and incentivize more clients to participate in the learning process at the same time.

References

1. Abadi, M., et al.: Deep learning with differential privacy. In: Proceedings of the 2016 ACM SIGSAC Conference on Computer and Communications Security, pp. 308–318 (2016)
2. Arachchige, P.C.M., Bertok, P., Khalil, I., Liu, D., Camtepe, S., Atiquzzaman, M.: Local differential privacy for deep learning. IEEE Internet Things J. **7**(7), 5827–5842 (2019)
3. Bach, S., Binder, A., Montavon, G., Klauschen, F., Müller, K.R., Samek, W.: On pixel-wise explanations for non-linear classifier decisions by layer-wise relevance propagation. PLoS ONE **10**(7), e0130140 (2015)
4. Dwork, C.: Differential privacy. In: Bugliesi, M., Preneel, B., Sassone, V., Wegener, I. (eds.) ICALP 2006. LNCS, vol. 4052, pp. 1–12. Springer, Heidelberg (2006). https://doi.org/10.1007/11787006_1
5. D'Acquisto, G., Domingo-Ferrer, J., Kikiras, P., Torra, V., de Montjoye, Y.A., Bourka, A.: Privacy by design in big data: an overview of privacy enhancing technologies in the era of big data analytics. arXiv preprint arXiv:1512.06000 (2015)
6. Dalenius, T.: Towards a methodology for statistical disclosure control. statistik Tidskrift **15**(429–444), 2–1 (1977)
7. Duchi, J.C., Jordan, M.I., Wainwright, M.J.: Local privacy and statistical minimax rates. In: 2013 IEEE 54th Annual Symposium on Foundations of Computer Science, pp. 429–438. IEEE (2013)
8. Dwork, C., McSherry, F., Nissim, K., Smith, A.: Calibrating noise to sensitivity in private data analysis. In: Halevi, S., Rabin, T. (eds.) TCC 2006. LNCS, vol. 3876, pp. 265–284. Springer, Heidelberg (2006). https://doi.org/10.1007/11681878_14
9. Geyer, R.C., Klein, T., Nabi, M.: Differentially private federated learning: a client level perspective. arXiv preprint arXiv:1712.07557 (2017)
10. Gong, N.Z., et al.: Joint link prediction and attribute inference using a social-attribute network. ACM Trans. Intell. Syst. Technol. (TIST) **5**(2), 1–20 (2014)
11. Ji, S., Li, W., Gong, N.Z., Mittal, P., Beyah, R.A.: On your social network de-anonymizablity: quantification and large scale evaluation with seed knowledge. In: NDSS (2015)
12. Ji, S., Wang, T., Chen, J., Li, W., Mittal, P., Beyah, R.: De-SAG: on the de-anonymization of structure-attribute graph data. IEEE Trans. Dependable Secure Comput. **16**, 594–607 (2017)
13. Jiang, H., Pei, J., Yu, D., Yu, J., Gong, B., Cheng, X.: Applications of differential privacy in social network analysis: a survey. IEEE Trans. Knowl. Data Eng. (2021)

14. Jiang, H., Yu, J., Cheng, X., Zhang, C., Gong, B., Yu, H.: Structure-attribute-based social network deanonymization with spectral graph partitioning. IEEE Trans. Comput. Soc. Syst. (2021)

15. Jiang, H., Yu, J., Hu, C., Zhang, C., Cheng, X.: Sa framework based deanonymization of social networks. Procedia Comput. Sci. **129**, 358–363 (2018)

16. Labitzke, S., Werling, F., Mittag, J., Hartenstein, H.: Do online social network friends still threaten my privacy? In: Proceedings of the Third ACM Conference on Data and Application Security and Privacy, pp. 13–24 (2013)

17. Lecuyer, M., Atlidakis, V., Geambasu, R., Hsu, D., Jana, S.: Certified robustness to adversarial examples with differential privacy. In: 2019 IEEE Symposium on Security and Privacy (SP), pp. 656–672. IEEE (2019)

18. Li, N., Li, T., Venkatasubramanian, S.: t-Closeness: privacy beyond k-anonymity and l-diversity. In: 2007 IEEE 23rd International Conference on Data Engineering, pp. 106–115. IEEE (2007)

19. Li, N., Qardaji, W., Su, D.: On sampling, anonymization, and differential privacy or, k-anonymization meets differential privacy. In: Proceedings of the 7th ACM Symposium on Information, Computer and Communications Security, pp. 32–33 (2012)

20. Machanavajjhala, A., Kifer, D., Gehrke, J., Venkitasubramaniam, M.: l-diversity: privacy beyond k-anonymity. ACM Trans. Knowl. Discovery Data (TKDD) **1**(1), 3-es (2007)

21. McSherry, F., Talwar, K.: Mechanism design via differential privacy. In: 48th Annual IEEE Symposium on Foundations of Computer Science (FOCS 2007), pp. 94–103. IEEE (2007)

22. Phan, N., Wu, X., Hu, H., Dou, D.: Adaptive Laplace mechanism: differential privacy preservation in deep learning. In: 2017 IEEE International Conference on Data Mining (ICDM), pp. 385–394. IEEE (2017)

23. Shirani, F., Garg, S., Erkip, E.: Optimal active social network de-anonymization using information thresholds. In: 2018 IEEE International Symposium on Information Theory (ISIT), pp. 1445–1449. IEEE (2018)

24. Wei, K., et al.: Federated learning with differential privacy: algorithms and performance analysis. IEEE Trans. Inf. Forensics Secur. **15**, 3454–3469 (2020)

25. Wei, W., Liu, L., Loper, M., Chow, K.H., Gursoy, M.E., Truex, S., Wu, Y.: A framework for evaluating gradient leakage attacks in federated learning. arXiv preprint arXiv:2004.10397 (2020)

Towards Building Intrusion Detection Systems for Multivariate Time-Series Data

ChangMin Seong[1], YoungRok Song[2], Jiwung Hyun[2], and Yun-Gyung Cheong[2(✉)]

[1] Department of Computer Software, Sungkyunkwan University, Suwon, South Korea
[2] Department of Artificial Intelligence, Sungkyunkwan University, Suwon, South Korea
ygcheong@gmail.com

Abstract. Recent network intrusion detection systems have employed machine learning and deep learning algorithms to defend against dynamically evolving network attacks. While most previous studies have focused on detecting attacks which can be determined based on a single time instant, few studies have paid attention to subsequence outliers, which require inspecting consecutive points in time for detection. To address this issue, this paper applies a time-series anomaly detection method in an unsupervised learning manner. To this end, we converted the UNSW-NB15 dataset into the time-series data. We carried out a preliminary evaluation to test the performance of the anomaly detection on the created time-series network dataset as well as on a time-series dataset obtained from sensors. We analyze and discuss the results.

Keywords: Time series · Intrusion detection system · Stacked RNN · Unsupervised learning · Anomaly detection

1 Introduction

Due to the rapid development and popularization of networks, security issues are also becoming an important issue. In order to solve these security issues, a network intrusion detection system (NIDS) has been widely used. A NIDS is a system that reads network packets and detects attack traffic and is known as an effective defense method against network security issues. During the last decade, network security systems have been developed by employing various time-series intrusion detection techniques. Pankaj et al. [21] propose a Long Short Term Memory Networks based Encoder-Decoder scheme for Anomaly Detection (EncDec-AD) that learns to reconstruct normal time-series behavior. Kyle et al. [22] demonstrate the effectiveness of LSTM and propose dynamic thresholding approach using LSTMs. Ding et al. [23] propose a real-time anomaly detection algorithm (RADM) based on Hierarchical Temporal Memory (HTM) and Bayesian Network (BN). Park et al. [24] introduced a long short-term memory-based variational autoencoder (LSTM-VAE) that fuses signals and reconstructs expected distribution.

Furthermore, unsupervised learning algorithms have been getting more attention owing to their advantage of training the models without labels during the training phase [11, 12]. In the unsupervised methods, attacks are generally detected by regarding them as outliers or anomalies. More details about outlier detection can be found in [1, 2, 10].

© The Author(s) 2022
S.-Y. Chang et al. (Eds.): SVCC 2021, CCIS 1536, pp. 45–56, 2022.
https://doi.org/10.1007/978-3-030-96057-5_4

Time-series data mean the data annotated with time stamps, collected at regular time intervals. Depending on what is considered an outlier, time-series outliers are largely divided into two types: point outliers and subsequence outliers [2]. A point outlier means an outlier of which value is significantly different from the values of the surrounding data in the overall flow of data in time order as shown in Fig. 1. In the figure, a point between 10 and 11 can be regarded as normal with a global perspective where similar data values exist between 21 and 22, but it is determined as an outlier considering the values of its neighbors with a local perspective [3]. These outliers can be determined relying on their characteristics at a specific time instant.

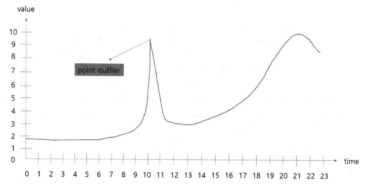

Fig. 1. An illustration of a point outlier where samples between 10 and 11 are spiking, distinguished from their neighboring data.

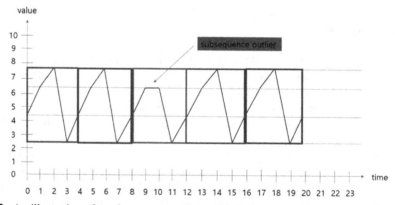

Fig. 2. An illustration of a subsequence outlier which is represented in the red box. The data values are within the minimum and the maximum of normal data, and yet the overall pattern is different from the rest. (Color figure online)

On the contrary, a subsequence outlier can be found only by inspecting consecutive instants in time. A subsequence outlier shows a pattern that deviates from the normal repetitive patents as shown in Fig. 2. The points between 9 and 10 can be regarded as normal when simply looking at the numerical values, but it is determined as an outlier

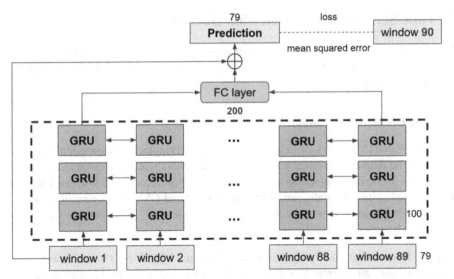

Fig. 3. The model structure uses stacked RNN(GRU) models. For the sliding window, which is the time interval the model trains the specific pattern, set to 90. Using the output of previous 89 data, the model predicts 90th data in the window. The numbers 79 denote the number of features excluding the time feature, 100 denotes the number of hidden cells of GRU, and 200 denotes the number of nodes of the FC (Fully Connected) layer.

since its pattern deviated from the repeating patterns between 1 and 2, 5 and 6, 13 and 14, and 17 and 18 [3]. Therefore, it is necessary to detect both outliers for building an intrusion detection system for practical domains. However, most previous studies have focused on detecting point outliers [6, 20].

To address this issue, this paper attempts to detect attacks using multivariate time-series network data. Since time-series network datasets are rarely available, we created a time-series network dataset using the UNSW-NB15 network dataset [7, 13–16]. As an experimental model, we employ an unsupervised approach which contains a stacked RNN model, as was provided by the DACON's HAICon2021 competition [17]. The approach showed a good performance, achieving F1 of 0.926 when the provided code was run on the HAI 2.0 dataset [4]. We carried out preliminary evaluations to test if this approach can be applied to the time-series network data.

2   Model

We use a stacked RNN (GRU) model [5] for learning time-series data in an unsupervised learning manner to detect attacks, which was provided as the baseline model for the HAICon2021 competition. This model uses a three-layer bidirectional GRU with 100 hidden cells as illustrated in Fig. 3. We use the experiment configuration that was set for the baseline model for comparison in the future research. We train the model for 32 epochs keeping the best model parameters, and the parameters that result in the best loss were chosen for evaluation. The window size was set as 90.

3 Time-Series Anomaly Detection Datasets

To evaluate the time-series anomaly detection system we selected two datasets, UNSW-NB15 dataset [7] and HAI 2.0 dataset [4]. The UNSW-NB15 dataset is converted into a time-series format.

3.1 The UNSW-NB15 Dataset

The UNSW-NB15 dataset is widely used for benchmarking network intrusion detection systems. The dataset contains 9 network attack behaviors which are Fuzzers, Analysis, Backdoors, DoS, Exploits, Generic, Reconnaissance, Shellcode, and Worms. The data are provided in two formats, raw traffic packet file and CSV file containing features extracted from captured network flows. We follow Ge et al. [8] to convert the packet data into a time-series format.

Feature Extraction: The raw traffic packets from the UNSW-NB15 dataset were captured using the IXIA PerfectStorm tool and are provided in the PCAP file format [7]. We first select and extract packet fields from the PCAP file using the TShark analyzer tool. Details of the selected fields are shown in Table 1.

Table 1. Detailed information of extracted fields from network packets.

Feature	Field detail
frame	frame.time_epoch, frame.len
ip	ip.src, ip.dst, ip.ttl
tcp	tcp.srcport, tcp.dstport, tcp.stream, tcp.len, tcp.checksum
udp	udp.srcport, udp.dstport, udp.stream, udp.checksum, udp.length

The UNSW-NB15 CSV file contains the flow-based features of labeled flow data. The description of 49 features in the file are listed in Table 2. Each flow is labelled as 0 for normal records and 1 for attacks.

Packet Labelling: After extracting the features from each packet, we sort them in the chronological order using the *frame.time_epoch* feature, which indicates the time information of the packet. The packets in the PCAP file are labelled using the labels in the CSV file. It has information about packets transmitted and a label denoting normal or attack. A label can be created by using the label feature value of the flow which contains the packet.

The process of determining whether a particular packet belongs to a flow is as follows. First, *frame.time_epoch* of the PCAP file is matched with the *Stime* value (the 29th field) and the *Ltime* value (the 30th field) of the CSV file. Among the data matched with the packet, we extracted the data that matches the *ip.src* and *ip.dst* of the PCAP with the

Table 2. Description of features.

Number	Description	Number	Description	Number	Description
1	srcip	18	Dpkts	35	ackdat
2	sport	19	swin	36	is_sm_ips_ports
3	dstip	20	dwin	37	ct_state_ttl
4	dsport	21	stcpb	38	ct_flw_http_mthd
5	proto	22	dtcpb	39	is_ftp_login
6	state	23	smeansz	40	ct_ftp_cmd
7	dur	24	dmeansz	41	ct_srv_src
8	sbytes	25	trans_depth	42	ct_srv_dst
9	dbytes	26	res_bdy_len	43	ct_dst_ltm
10	sttl	27	Sjit	44	ct_src_ ltm
11	dttl	28	Djit	45	ct_src_dport_ltm
12	sloss	29	Stime	46	ct_dst_sport_ltm
13	dloss	30	Ltime	47	ct_dst_src_ltm
14	service	31	Sintpkt	48	attack_cat
15	Sload	32	Dintpkt	49	Label
16	Dload	33	tcprtt		
17	Spkts	34	synack		

first field *srcip* and the third field *dstip* of the CSV file. Finally, for TCP, we matched *tcp.srcport* and *tcp.dstport* in the PCAP file, and in the case of UDP, *udp.srcport* and *udp.dstport* in the PCAP file with the 2nd field *sport*, and 4th field *dsport* of the CSV file, and the label of the matched file becomes the label of the corresponding PCAP file. If there is no matching data, it is infeasible to determine whether it is normal or an attack, hence, we removed the corresponding packet. *Tcp* information and *udp* information are integrated into one common information, and then in the case of *ip.src* and *ip.dst*, they are used up to map the PCAP file and the CSV information and then removed. Finally, in the created time-series network data, there are 9 features: *frame.time_epoch*, *frame.len*, *ip.ttl*, *srcport*, *dstport*, *stream*, *checksum*, *len*, and *label*. We removed the label from the data for train, validation and test, since we apply unsupervised learning to dataset, we only used the label for evaluation for validation and test. In total, there are 295,342 time-series data with 277,828 normal data and 17,514 attack data.

Preprocessing: For the source port and destination port features, the port numbers greater than 49,152 are labelled as 2, the numbers greater than 1,024 are labelled to 1, and the numbers lower than 1,024 are labelled to 0 since they are divided to dynamic port, registered port and well-known port. Then numerical features were scaled to fit 0 to 1 using a min-max scaler.

3.2   The HAI 2.0 Dataset

The HAI 2.0 dataset is a time-series dataset created for attack detection in cyber-physical systems such as railways, water-treatment, and power plants [4]. The data were collected from the four processes: the boiler process, the turbine process, the water-treatment process, and the HIL simulation. Data samples were collected every second and consist of 80 features. Normal data were collected for 7 continuous days, and the attack data include 38 different attack types. The data are sorted in the increasing order of time feature in the format of "yyyy-MM-dd hh:mm:ss.". Other features contain information associated with the processes such as temperature setpoint, water level setpoint and motor speed.

Preprocessing: To preprocess the data, the timestamp features were dropped, and the numerical features were scaled with a min-max scaler similar to UNSW-NB15 [17]. For some features, of which maximum value and minimum value are the same, we set these features as 0. After scaling features, we applied an exponential weighted function in python function "ewm" with 0.9 for alpha for noise smoothing.

4   Experiments

We compare and analyze the anomaly detection system performance using the UNSW-NB15 and the HAI 2.0 dataset. We convert attack detection into an anomaly detection problem by assuming the attack to be anomalous.

4.1   Data Preparation

For both datasets, an unsupervised learning was conducted to train the model using only normal data. We divided the time-series network dataset into training, validation, and test datasets in a ratio of 8:1:1. Then, since the attack data is also included in the training datasets for the time-series network data, we removed attack data in the training datasets. The number of instances for each dataset is presented in Table 3.

Table 3. Simple statistics of processed UNSW-NB15 dataset.

	Training	Validation	Test
Normal	226,240	25,706	25,882
Attack	0	3,828	3,652
Total	226,240	29,534	29,534

However, there are no labels in the test dataset of HAI 2.0 dataset. For the evaluation, we divided the validation dataset, which has labels, into the validation dataset (first 50%) and the test dataset (last 50%). Table 4 shows the simple statistics of the processed dataset.

Table 4. Simple statistics of the processed HAI 2.0 dataset.

	Training	Validation	Test
Normal	965,603	21,060	21,512
Attack	0	540	89
Total	965,603	21,600	21,601

4.2 Training

As described in Fig. 3, the model is trained to predict the last sample in the given time window when the preceding samples are given. In order to predict whether the last sample is an anomaly the model is only trained with windows containing normal samples. Theoretically the model will predict the last sample as close as possible to the normal sample given the preceding sample. Therefore, if the difference between the prediction and true last sample is significant, we consider the last sample to be an anomaly. We predict the last sample of the window as an anomaly if the difference is greater than a predetermined threshold. The parameters for training the model are provided in Table 5. The stride means how much data to skip during training.

Table 5. Model parameters and configurations.

Parameter	Value/Name	Parameter	Value/Name
n_hidden	100	n_layers	3
batch_size	512	num_epochs	32
window_size	90	stride	10
loss	MSE	optimizer	AdamW
scheduler	X	dropout	X

4.3 The Evaluation Metrics

There are various evaluation metrics such as precision, recall, and $F1$ that are frequently used. However, the evaluation metric of time-series data needs to consider various factors such as the diversity of detected attacks and the accuracy of detection as illustrated in Fig. 4.

For example, as shown in Fig. 4, Model 2 detects 3 anomaly instances between 0 and 3, and Model 1 detects 2 instances, one between 1 and 2 and the other between 6 and 7. In terms of accuracy, Model 2 outperforms Model 1. However, considering that Model 2 does not detect anomalies between 6 and 8 time slots, it is hard to determine which model performs better. TaPR [19] is an evaluation metric that considers these factors. TaP, which corresponds to precision, is an evaluation metric indicating whether

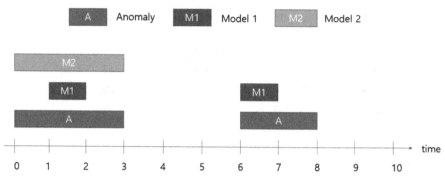

Fig. 4. Illustration of time-series anomaly detection where the two different models Model 1 and Model 2 are used, modified from [18]. The X-axis indicates time, and *A* indicates the time slots where an anomaly exists. *M1* indicates the anomalies that Model 1 detects, and *M2* indicates the anomalies that Model 2 detects.

the prediction finds outliers with less false positives. TaR, which corresponds to recall, is an evaluation metric indicating the diversity of the anomalies. Using the detection score TaP^d (resp. TaR^d) and the portion score TaP^p (resp. TaR^p), TaP and TaR can be calculated as follows:

$$TaP = \alpha \times TaP^d + (1 - \alpha) \times TaP^p \tag{1}$$

$$TaR = \alpha \times TaR^d + (1 - \alpha) \times TaR^p \tag{2}$$

where α controls the ratio of TaP^d (resp. TaR^d) and TaP^p (resp. TaR^p), and its value is between 0 and 1 [9].

5   The Experiment Results

This section reports the evaluation results. The figures below show the error and attack distribution of the time-series network data created in this paper and the HAI 2.0 data, respectively (Figs. 5 and 6).

Using the experimental results of validation data, the threshold was set to 0.04 for the HAI 2.0 data, and the threshold was set to 0.2 for time-series network data. The two dataset show different properties. In the HAI 2.0 data, the attack data tends to be greater than the normal data, while in the time-series network data values are relatively evenly distributed. In addition, in the case of the HAI 2.0 dataset, the number of normal data is overwhelmingly larger than that of attack data, unlike the time-series network data. As the evaluation metric, we use TaPR described in Sect. 4.3. The analyses of the results are shown in the following Tables 6 and 7.

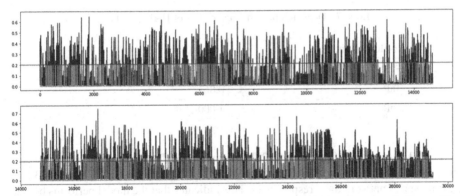

Fig. 5. Distribution of error and attack in validation dataset of the time-series network dataset. The x-axis indicates the order of the data, and the y-axis indicates the absolute difference of (answer - guess). The orange line indicates the attack position, and the blue line indicates the size of the error. The red line is the threshold value that separates the boundary between normal and attack. (Color figure online)

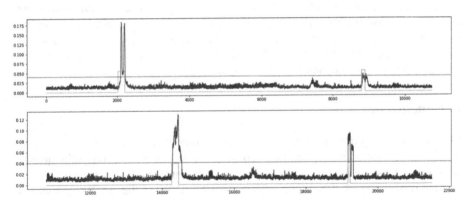

Fig. 6. Distribution of error and attack in validation dataset of the HAI 2.0 dataset. The x-axis indicates the order of the data, and the y-axis indicates the absolute difference of (answer - guess). The orange line indicates the attack position, and the blue line indicates the size of the error. The red line is the threshold value that separates the boundary between normal and attack. (Color figure online)

Table 6. Detection performance results of UNSW-NB15 data.

Evaluation metric	UNSW-NB15 data
F1	0.737
TaP	0.731
TaR	0.743

Table 7. Detection performance results of HAI 2.0 data.

Evaluation metric	HAI 2.0 data
F1	0.926
TaP	0.861
TaR	1.000

The F1 scores are 0.926 for HAI 2.0 data and 0.737 for time-series network data. The TaP and TaR scores are 0.861 and 1.000 for HAI 2.0 data, and 0.731 and 0.743 for the time-series network data, respectively. This indicates that the model performs better with the HAI 2.0 dataset which contains sensor data.

There are two main factors that account for the poor performance of the time-series network dataset. First, the number of features in the time-series network dataset may be insufficient. In the case of the HAI 2.0 dataset, there are about 80 features, in the case of the time-series network data, only about 10 features were used, making it difficult to determine its anomaly. The other reason is that time-series network data are not complete time-series. In the case of HAI 2.0 dataset, data is generated every second, but in the case of the time-series network dataset, since packets are not transmitted at a specific period, it is difficult to generate data at regular intervals. Moreover, since the attack data is removed from the training data of the time-series network data to learn the normal data only, the time information becomes more irregular.

6 Conclusion

While unsupervised deep learning models have shown great performances in detecting attacks that are point outliers, little has been researched on detecting subsequence outliers. For building a NIDS which can detect subsequence outliers, we first created the time-series network data by processing the UNSW-NB15 dataset. We carried out preliminary experiments using both the HAI 2.0 dataset and the time-series network dataset we created, using a stacked RNN model in an unsupervised manner. The results show that the model performs better with run on the HAI 2.0 dataset than tested on the time-series network dataset. The model achieved F1 scores of 0.926 for the HAI 2.0 data and 0.737 for the time-series network data. The TaP and TaR scores are 0.861 and 1.000 for the HAI 2.0 data, and 0.731 and 0.743 for the time-series network data. The lack of data and insufficient features of the time-series network data can account for its poor performance. We expect that more studies on time-series network data attack detection in the future will help solve these shortcomings.

Acknowledgement. This work was supported by Institute of Information & communications Technology Planning & Evaluation (IITP) grant funded by the Korea government (MSIT) (No. 2020-0-00952, Development of 5G Edge Security Technology for Ensuring 5G+ Service Stability and Availability).

References

1. Braei, M., Wagner, S.: Anomaly detection in univariate time-series: a survey on the state-of-the-art. ArXiv abs/2004.00433 (2020)
2. Bl'azquez-Garc'ia, A., et al.: A review on outlier/anomaly detection in time series data. ACM Comput. Surv. (CSUR) **54**, 1–33 (2021)
3. "Anomaly Detection in Time Series: 2021", neptune.ai. 19 July 2021. https://neptune.ai/blog/anomaly-detection-in-time-series. Accessed 5 Sept 2021
4. Shin, H.-K., Lee, W., Yun, J.-H., Kim, H.: HAI 1.0: HIL-based augmented ICS security dataset. In: 13th USENIX Workshop on Cyber Security Experimentation and Test (2020)
5. Cho, K., et al.: Learning phrase representations using RNN encoder–decoder for statistical machine translation. In: EMNLP (2014)
6. Sandosh, S., Govindasamy, V., Akila, G.: Enhanced intrusion detection system via agent clustering and classification based on outlier detection. Peer-to-Peer Network. Appl. **13**(3), 1038–1045 (2020). https://doi.org/10.1007/s12083-019-00822-3
7. Moustafa, N., Slay, J.:UNSW-NB15: a comprehensive data set for network intrusion detection systems (UNSW-NB15 network data set). In: Military Communications and Information Systems Conference (MilCIS). IEEE (2015)
8. Ge, M., et al.: Deep learning-based intrusion detection for IoT networks. In: 2019 IEEE 24th Pacific Rim International Symposium on Dependable Computing (PRDC), pp. 256–25609 (2019)
9. Hwang, W.-S., Yun, J.-H., Kim, J., Kim, H.: Time-series aware precision and recall for anomaly detection: considering variety of detection result and addressing ambiguous labeling, pp. 2241–2244 (2019). https://doi.org/10.1145/3357384.3358118
10. Gupta, M., Gao, J., Aggarwal, C.C., Han, J.: Outlier detection for temporal data: a survey. IEEE Trans. Knowl. Data Eng. **26**(9), 2250–2267 (2014). https://doi.org/10.1109/TKDE.2013.184
11. Song, Y., Hyun, S., Cheong, Y.-G.: A systematic approach to building autoencoders for intrusion detection. In: Park, Y., Jadav, D., Austin, T. (eds.) SVCC 2020. CCIS, vol. 1383, pp. 188–204. Springer, Cham (2021). https://doi.org/10.1007/978-3-030-72725-3_14
12. Song, Y., Hyun, S., Cheong, Y.-G.: Analysis of autoencoders for network intrusion detection. Sensors **21**(13), 4294 (2021). https://doi.org/10.3390/s21134294
13. Moustafa, N., Slay, J.: The evaluation of network anomaly detection systems: statistical analysis of the UNSW-NB15 dataset and the comparison with the KDD99 dataset. Inf. Secur. J. Glob. Perspect., 1–14 (2016)
14. Moustafa, N., et al.: Novel geometric area analysis technique for anomaly detection using trapezoidal area estimation on large-scale networks. IEEE Trans. Big Data (2017)
15. Moustafa, N., Creech, G., Slay, J.: Big data analytics for intrusion detection system: statistical decision-making using finite dirichlet mixture models. In: Palomares Carrascosa, I., Kalutarage, H. K., Huang, Y. (eds.) Data Analytics and Decision Support for Cybersecurity. DA, pp. 127–156. Springer, Cham (2017). https://doi.org/10.1007/978-3-319-59439-2_5
16. Sarhan, M., Layeghy, S., Moustafa, N., Portmann, M.: NetFlow datasets for machine learning-based network intrusion detection systems. In: Deze, Z., Huang, H., Hou, R., Rho, S., Chilamkurti, N. (eds.) BDTA/WiCON -2020. LNICSSITE, vol. 371, pp. 117–135. Springer, Cham (2021). https://doi.org/10.1007/978-3-030-72802-1_9
17. "HAI DataSet Baseline Model", DACON, 2 August 2021. https://dacon.io/competitions/official/235757/codeshare/3009?page=1&dtype=recent. Accessed 5 Sept 2021
18. "[Paper Review] Evaluation Metrics for Time Series Anomaly Detection", DSBA, 23 September 2020. http://dsba.korea.ac.kr/seminar/?pageid=3&mod=document&uid=1332. Accessed 6 Sept 2021

19. Hwang, W.-s., Yun, J.-H., Kim, J., Kim, H.: Time-series aware precision and recall for anomaly detection - considering variety of detection result and addressing ambiguous labeling. In: CIKM 2019: Proceedings of the 28th ACM International Conference on Information and Knowledge Management (2019)

20. Devan, P., Khare, N.: An efficient XGBoost–DNN-based classification model for network intrusion detection system. Neural Comput. Appl. **32**(16), 12499–12514 (2020). https://doi.org/10.1007/s00521-020-04708-x

21. Malhotra, P., et al.: LSTM-based encoder-decoder for multi-sensor anomaly detection. arXiv preprint arXiv:1607.00148 (2016)

22. Hundman, K., et al.: Detecting spacecraft anomalies using lstms and nonparametric dynamic thresholding. In: Proceedings of the 24th ACM SIGKDD International Conference on Knowledge Discovery & Data Mining (2018)

23. Ding, N., et al.: Multivariate-time-series-driven real-time anomaly detection based on bayesian network. Sensors **18**(10), 3367 (2018)

24. Park, D., Hoshi, Y., Kemp, C.C.: A multimodal anomaly detector for robot-assisted feeding using an lstm-based variational autoencoder. IEEE Rob. Autom. Lett. **3**(3), 1544–1551 (2018)

Encryption

Encryption Scheme Based on the Generalized Suzuki 2-groups and Homomorphic Encryption

Gennady Khalimov[1] ⓘ, Yevgen Kotukh[2(✉)] ⓘ, Sang-Yoon Chang[3] ⓘ,
Yaroslav Balytskyi[3] ⓘ, Maksym Kolisnyk[1] ⓘ, Svitlana Khalimova[1] ⓘ,
and Oleksandr Marukhnenko[1] ⓘ

[1] Kharkiv National University of Radioelectronics, Kharkiv, Ukraine
[2] Sumy State University, Sumy, Ukraine
[3] University of Colorado Colorado Springs, Colorado Springs, CO, USA

Abstract. This article describes a new implementation of MST-based encryption for generalized Suzuki 2-groups. The well-known MST cryptosystem based on Suzuki groups is built on a logarithmic signature at the center of the group, resulting in a large array of logarithmic signatures. An encryption scheme based on multiparameter non-commutative groups is proposed. The multiparameter generalized 2 - Suzuki group was chosen as one of the group constructions. In this case, a logarithmic signature is established for the entire group. The main difference from the known one is the use of homomorphic encryption to construct coverings of logarithmic signatures for all group parameters. This design improves a secrecy of the cryptosystem is ensured at the level of a brute-force attack.

Keywords: MST cryptosystem · Logarithmic signature · Random cover · Generalized Suzuki 2-groups

1 Introduction

Recent advances in quantum computing for solving complex problems formulate new trends for building secure public-key cryptosystems. The main directions in this area are the solution of the problem of finding the conjugate element in the theory of non-commutative groups and the word problem in groups and semigroups. The word complexity problem was proposed by Wagner and Magyarik [1] and implemented in several cryptosystems. One of the best known and most studied is a cryptosystem based on factorization in finite groups of permutations, called the logarithmic signature [2]. In 2009, Lempken et al. described an MST3 public-key cryptosystem based on a logarithmic signature and a Suzuki 2-group [2]. In 2008 Magliveras et al. [4] presented a comprehensive analysis of the MST3 cryptosystem identifying limitations for the logarithmic signature and stated that the transitive logarithmic signature is not suitable for the MST3 cryptosystem. In 2010, Swaba et al. [5] analyzed all known attacks on MST cryptography and built a more secure eMST3 cryptosystem by adding a secret homomorphic coverage. In 2018, T. van Trung [7] proposed a general method for constructing strong

© The Author(s) 2022
S.-Y. Chang et al. (Eds.): SVCC 2021, CCIS 1536, pp. 59–76, 2022.
https://doi.org/10.1007/978-3-030-96057-5_5

aperiodic logarithmic signatures for Abelian p-groups, which is a further contribution to the practical application of MST cryptosystems.

The construction of MST cryptosystems based on multiparameter non-commutative groups was proposed in [7–9]. MST cryptosystems based on multi-parameter groups allow optimizing the costs of cryptosystem parameters and secrecy.

Generalized Suzuki 2-groups are multivariable and have the highest group order compared to other multivariable groups. The first implementation of the cryptosystem on the generalized Suzuki 2-group is presented in [8] and does not provide protection against brute force attacks with sequential brute force key recovery. Analysis of MST cryptosystems by group shows their vulnerability to highlighted text attacks. The design feature of all known MST implementations is the presence of known texts and, as a consequence, the possibility of such cryptanalysis. A secure encryption scheme is proposed based on the generic Suzuki 2-group with homomorphic encryption.

2 Proposal

The generalizations of Suzuki 2-groups is defined over a finite field, F_q, $q = 2^n$, $n > 0$ for a positive integer l and $a_1, a_2, ..., a_l \in F$ for some automorphism θ of F as [10]:

$$A_l(n, \theta) = \{ S(a_1, a_2, ..., a_l) | a_i \in F_q \}$$

Each element of $A_l(n, \theta)$ can be expressed uniquely and it follows that $|A_l(n, \theta)| = 2^{nl}$ and $A_l(n, \theta)$ define a group of order 2^{nl}. If $l = 2$, this group is isomorphic to a Suzuki 2-group $A(n, \theta)$.

Group operation is defined as a product:

$$S(a_1, a_2, ..., a_l) S(b_1, b_2, ..., b_l) = S(a_1 + b_1, a_2 + (a_1\theta)b_1$$
$$+ b_2, a_3 + (a_2\theta)b_1 + (a_1\theta^2)b_2 + b_3,$$
$$..., a_l + (a_{l-1}\theta)b_1 + ... + (a_1\theta^{l-1})b_{l-1} + b_l).$$

with the Identity element being $S(0_1, 0, ..., 0)$.
The inverse element is given by:

$$S(a_1, a_2, a_3, ..., a_l)^{-1} = S(a_1, a_2 + a_1\theta a_1, a_3 + a_2\theta a_1$$
$$+ a_1\theta^2(a_2 + a_1\theta a_1), ..., a_l + a_{l-1}\theta a_1 + ...).$$

The group G is nonabelian group and has nontrivial center:

$$Z(G) = \{ S(0, 0, ..., c) | c \in F_q \}.$$

Assume that θ is the Frobenius automorphism of F, $\theta : x \to x^2$. For the fixed finite field, the group $A_l(n, \theta)$ order is greater than the classical Suzuki 2 - group.

In the new implementation of the cryptosystem, we have changed the encryption algorithm and suggest using homomorphic encryption for random covers. In this case, the complexity of the key recovery attack will be determined by exhaustive search over the entire group.

2.1 Description of the Scheme

Our proposal is to create a logarithmic signature for the whole generalized Suzuki 2-group and homomorphic encryption of random covers in the logarithmic signature.

Let's take a look at the basic steps of encryption.

Key Generation.

We fix a large group $A_l(n, \theta) = \{S(a_1, a_2, ..., a_l) | a_i \in F_q\}, q = 2^n$.

Let's build a tame logarithmic signatures $\beta_k = [B_{1(k)}, ..., B_{s(k)}] = (b_{ij})_k = S(0, ..., 0, b_{ij(k)}, 0, ..., 0)$ of type: $(r_{1(k)}, ..., r_{s(k)}), i = \overline{0, s(k)}, j = \overline{1, r_{i(k)}}, b_{ij(k)} \in F_q, k = \overline{1, l}$.

Let's set a random cover:

$$\alpha_k = [A_{1(k)}, ..., A_{s(k)}] = (a_{ij})_k = S\left(a_{ij(k)}^{(1)}, a_{ij(k)}^{(2)}, ..., a_{ij(k)}^{(l)}\right)$$

of the same type as β_k, where $a_{ij} \in A_l(n, \theta), a_{ij(k)}^{(v)} \in F_q \backslash \{0\}, i = \overline{1, s}, j = \overline{1, r_{i(k)}}, k = \overline{1, l}$.

Select the random covers:

$w_{(k)} = [W_{1(k)}, ..., W_{s(k)}] = (w_{ij})_{(k)} = S\left(w_{ij(k)}^{(1)}, w_{ij(k)}^{(2)}, ..., w_{ij(k)}^{(l)}\right)$ of the same types as $\beta_{(k)}$, where $w_{ij} \in A_l(n, \theta), w_{ij(k)} \in F_q \backslash \{0\}, i = \overline{0, s(k)}, j = \overline{1, r_{i(k)}}, k = \overline{1, l}$.

Let's generate random $t_{0(k)}, ..., t_{s(k)} \in A_l(n, \theta) \backslash Z, t_{i(k)} = S(t_{i1(k)}, ..., t_{il(k)}), t_{ij(k)} \in F^\times, i = \overline{0, s(k)}, k = \overline{1, l}$. Choose

$$\tau_{0(k)}, ..., \tau_{s(k)} \in A_l(n, \theta) \backslash Z, \tau_{i(k)}$$
$$= S(\tau_{i1(k)}, ..., \tau_{il(k)}), \tau_{ij(k)} \in F^\times, i = \overline{0, s(k)}, k = \overline{1, l}.$$

Let's take $t_{s(k-1)} = t_{0(k)}, \tau_{s(k-1)} = \tau_{0(k)}, k = \overline{1, l}$.

Let's define an additional group operation:

$$S(a_1, a_2, ..., a_l) \circ^{(k)} S(b_1, b_2, ..., b_l) =$$
$$S(a_1 + b_1, a_2 + b_2, ..., a_k + b_k, a_{k+1} + a_k^2 b_1 + ... + a_1^{2^k} b_k$$
$$+ b_{k+1}, ..., a_l + a_{l-1}^2 b_1 + ... + a_1^{2^{l-1}} b_{l-1} + b_l).$$

The inverse element $S^{-(k)}$ for the group operation $\circ^{(k)}$ is

$$S^{-(k)}(a_1, a_2, ..., a_l) = S(a_1, a_2, ..., a_k, \alpha_{k+1}, ..., \alpha_l)$$

where

$$\alpha_{k+1} = a_{k+1} + a_k^2 a_1 + ... + a_2^{2^{k-1}} a_{k-1} + a_1^{2^k} a_k,$$
$$\alpha_{k+2} = a_{k+2} + a_{k+1}^2 a_1 + ... + a_3^{2^{k-1}} a_{k-1} + a_2^{2^k} a_k + a_1^{2^{k+1}} \alpha_{k+1},$$
$$...$$
$$\alpha_l = a_l + a_{l-1}^2 a_1 + ... + a_{l-k}^{2^k} a_k + a_{l-k-1}^{2^{k+1}} \alpha_{k+1} + , ..., + a_l^{2^{l-1}} \alpha_{l-1}$$

The application of additional group operation $\circ^{(k)}$ leads to homomorphic representation of group elements $S(a_1, a_2, ..., a_l) \xrightarrow{\circ^{(k)}} S(a_1, a_2, ..., a_k, \alpha_{k+1}, ..., \alpha_l) = S^{(k)}$.

We apply inverse homomorphic transformation for the inverse and direct elements $S_1^{-(k)}$, $S_2^{(k)}$ of the group for the calculation in group with left inverse element $S_1^{-(n)\circ}$.

$S_3 = S_1^{-(k)\circ} \cdot S_2^{(k)\circ}$ For $S_1^{-(k)}$ we have:

$$S^{-(k)\circ} = S^\circ(a_1, a_2, ..., a_k, \alpha_{k+1}, ..., \alpha_l) = S(\alpha_1, ..., \alpha_k, \alpha_{k+1}, ..., \alpha_l), \text{ where}$$

$$\alpha_1 = a_1, \alpha_2 = a_2 + a_1^2 a_1, ... \alpha_k = a_k + a_{k-1}^2 a_1 + ..., a_l^{2^{k-1}} a_{k-1}.$$

and for $S_2^{(k)}$ respectively to $S_3 = S_1^{-(k)\circ} \cdot S_2^{(k)\circ}$ we get

$$S^{(k)\circ} = S^\circ(b_1, b_2, ..., b_k, \beta_{k+1}, ..., \beta_l) = S(\beta_1, ..., \beta_k, \beta_{k+1}, ..., \beta_l)$$

$$\beta_1 = b_1, \beta_2 = b_2 + a_1^2(b_1 + a_1), ...$$
$$\beta_k = b_k + a_{k-1}^2(b_1 + a_1) + ..., a_l^{2^{k-1}}(b_{k-1} + a_{k-1}).$$

Homomorphic transformations for $S^{-(k)\circ}$, $S^{(k)\circ}$ are needed to for not breaking the group operation when calculating the elements of the group $A_l(n, \theta)$.

Let $f(e)$ be a homomorphic cryptographic transformation with respect to addition $f(a + b) = f(a) + f(b)$, $e, a, b \in F_q$ and the corresponding inverse transformation $\hat{f}(e) = e$. We calculate the covering of the logarithmic signatures:

$$h_{(k)} = \left[h_{1(k)}, ..., h_{s(k)}\right] = t_{(i-1)(k)}^{-(k)} \, o^{(k)} \, \left(w_{ij}\right)_{(k)} o^{(k)} \left(b_{ij}\right)_{(k)} o^{(k)} t_{i(k)}$$

and coverings of the homomorphic cryptographic transformation:

$$g_{(k)} = \left[g_{1(k)}, ..., g_{s(k)}\right] = \tau_{(i-1)(k)}^{-(k)} \, o^{(k)} f\left(w_{ij}\right)_{(k)} o^{(k)} \tau_{i(k)}, \text{ where}$$

$$f\left(w_{(k)}\right) = f\left(w_{ij}\right)_{(k)} = S\left(f\left(w_{ij(k)_1}\right), f\left(w_{ij(k)_2}\right), ..., f\left(w_{ij(k)_l}\right)\right),$$

$$i = \overline{1, s(k)}, \, j = \overline{1, r_{i(k)}}, \, k = \overline{1, l}.$$

An output public key is (a_k, h_k, g_k), and a private key $\left[f, \beta_{(k)}, \left(t_{0(k)}, ..., t_{s(k)}\right), \left(\tau_{0(k)}, ..., \tau_{s(k)}\right)\right]$, $k = \overline{1, l}$ respectively.

Encryption

Let the message to be $x = S(x_1, ..., x_l)$ and the public key (a_k, h_k, g_k), $k = \overline{1, l}$ respectively. Choose a random $R = (R_1, ..., R_l)$, $R_1, ..., R_l \in \mathbb{Z}_{|F_q|}$.

Compute the ciphertext y_1, y_2, y_3 as:

$$y_1 = \alpha(R) \cdot x = \alpha_1(R_1) \cdot \alpha_2(R_2) ... \alpha_l(R_l) \cdot x$$

$$= S\left(\sum_{k=1}^{l} \sum_{i=1, j=R_{i(k)}}^{s(k)} a_{ij(k)}^{(1)} + x_1, \sum_{k=1}^{l} \sum_{i=1, j=R_{i(k)}}^{s(k)} a_{ij(k)}^{(2)} + x_2 + *, \right.$$

$$\left. ..., \sum_{k=1}^{l} \sum_{i=1, j=R_{i(k)}}^{s(k)} a_{ij(k)}^{(l)} + x_l + *, \right),$$

$$y_2 = h(R) = h_1(R_1) o^{(1)}$$

$$\left(h_2(R_2) \, o^{(2)} \dots \left(h_{l-1}(R_{l-1}) \, o^{(l-2)} \left(h_{l-1}(R_{l-1}) \, o^{(l-1)} \, h_l(R_l) \right) \right) \right)$$

$$= S \left(\sum_{k=1}^{l} \sum_{i=1, j=R_{i(k)}}^{s(k)} w_{ij(k)}^{(1)} + \sum_{i=1, j=R_{i(1)}}^{s(1)} \beta_{ij(1)}, \sum_{k=1}^{l} \sum_{i=1, j=R_{i(k)}}^{s(k)} w_{ij(k)}^{(2)} \right.$$

$$\left. + \sum_{i=1, j=R_{i(2)}}^{s(2)} \beta_{ij(2)} + *, \dots, \sum_{k=1}^{l} \sum_{i=1, j=R_{i(k)}}^{s(k)} w_{ij(k)}^{(l)} + \sum_{i=1, j=R_{i(l)}}^{s(l)} \beta_{ij(l)} + * \right)$$

Here, the $(*)$ components are determined by cross-calculations in the group operation of the product of $t_{0(k)}, \dots, t_{s(k)}$ and the product of $w_{(k)}(R_k) + \beta_{(k)}(R_k)$.

$$y_3 = g(R) = g_1(R_1) o^{(1)}$$

$$\left(g_2(R_2) \, o^{(2)} \dots \left(g_{l-1}(R_{l-1}) \, o^{(l-2)} \left(g_{l-1}(R_{l-1}) \, o^{(l-1)} \, g_l(R_l) \right) \right) \right)$$

$$= S \left(\sum_{k=1}^{l} \sum_{i=1, j=R_{i(k)}}^{s(k)} f\left(w_{ij(k)}^{(1)} \right) +, \sum_{k=1}^{l} \sum_{i=1, j=R_{i(k)}}^{s(k)} f\left(w_{ij(k)}^{(2)} \right) + *, \dots, \right.$$

$$\left. \sum_{k=1}^{l} \sum_{i=1, j=R_{i(k)}}^{s(k)} f\left(w_{ij(k)}^{(l)} \right) + * \right)$$

Here, the $(*)$ components are determined by cross-calculations in the group operation of the product of $\tau_{0(k)}, \dots, \tau_{s(k)}$ and the product of $f\left(w_{(k)}(R_k) \right)$.

Output: a ciphertext (y_1, y_2, y_3) of the message x.

Decryption *Input*: a ciphertext (y_1, y_2, y_3) and a private key $\left[f, \beta_{(k)}, t_{i(k)}, \tau_{i(k)} \right]$, $i = \overline{0, s(k)}, k = \overline{1, l}$.

To decrypt a message x, we need to restore random numbers $R = (R_1, R_2, \dots, R_l)$. Compute

$$D^{(1)}(R) = D^{(1)}(R_1, R_2, \dots, R_l) = t_{0(1)} \, o^{(1)} \, y_2 \, o^{(l)} \, t_{s(l)}^{-(l)}$$

$$= S\left(\sum_{i=1, j=R_{i(1)}}^{s(1)} w_{ij(1)}^{(1)} + \beta_1(R_1), *, \dots, * \right),$$

$$G^{(1)}(R) = G^{(1)}(R_1, R_2, \dots, R_l) = \tau_{0(1)} \, o^{(1)} \, y_3 \, o^{(l)} \, \tau_{s(l)}^{-(l)}$$

$$= S\left(\sum_{i=1, j=R_{i(1)}}^{s(1)} f\left(w_{ij(1)}^{(1)} \right), *, \dots, * \right),$$

$$D^{(1)}(R)' = D^{(1)}(R) \, o^{(1)} \, \hat{f}(G^{(1)}(R))^{-(1)} = S\left(\sum_{i=1, j=R_{i(1)}}^{s(1)} \beta_{ij(1)}, *, * \right) \text{ Restore } R_1 \text{ with}$$

$$\beta_{(1)}(R_1) = \sum_{i=1, j=R_{i(1)}}^{s(1)} \beta_{ij(1)} \text{ using } \beta_{(1)}(R_1)^{-1}, \text{ because } \beta_1 \text{ is simple.}$$

For the further calculation, it is necessary to remove the component $h_1(R_1)$ from y_2 and $g_1(R_1)$ from y_3. Compute

$$y_2^{(1)} = h_1(R_1)^{-(1)\circ} \cdot y_2^\circ, \ y_3^{(1)} = g_1(R_1)^{-(1)\circ} \cdot y_3^\circ, \ D(R)^{(2)} = t_{0(2)} \circ^{(2)} y_2^{(1)} \circ^{(l)} t_{s(l)}^{-(l)},$$

$$G(R)^{(2)} = t_{0(2)} \circ^{(2)} y_3^{(1)} \circ^{(l)} t_{s(l)}^{-(l)},$$

$$D^{(2)}(R)' = D^{(2)}(R) \circ^{(2)} \hat{f}(G^{(2)}(R))^{-(2)} = S(0, \sum_{i=1, j=R_{i(2)}}^{s(2)} \beta_{ij(2)_c}, *).$$

and restore R_2 with $\beta_{(2)}(R_2) = \sum_{i=1, j=R_{i(2)}}^{s(2)} \beta_{ij(2)}$ using $\beta_{(2)}(R_2)^{-1}$, because β_2 is simple. We continue the calculations iteratively until the last value R_l is restored. We have the following recurrent relations for $n = 1, l-1$:

$$y_2^{(n)} = h_n(R_n)^{-(n)\circ} \cdot y_2^{(n-1)\circ}, y_3^{(n)} = g_n(R_n)^{-(n)\circ} \cdot y_3^{(n-1)\circ},$$

$$D^{(n+1)}(R) = t_{0(n+1)} \circ^{(n+1)} y_2^{(n)} \circ^{(l)} t_{s(l)}^{-(l)}, G^{(n+1)}(R) = t_{0(n+1)} \circ^{(n+1)} y_3^{(n)} \circ^{(l)} t_{s(l)}^{-(l)},$$

$$D^{(n+1)}(R)' \quad = \quad D^{(n+1)}(R) \quad \circ^{(n+1)} \quad \hat{f}(G^{(n+1)}(R))^{-(n+1)} \quad =$$

$$S(0, 0, ..., 0, \sum_{i=1, j=R_{i(n+1)}}^{s(n+1)} \beta_{ij(n+1)}, *)$$

Restore R_{n+1} with $\beta_{(n+1)}(R_{n+1}) = \sum_{i=1, j=R_{i(n+1)}}^{s(n+1)} \beta_{ij(n+1)}$ using $\beta_{(n+1)}(R_{n+1})^{-1}$.

Recovery of the message $x = a(R_1, R_2, ..., R_l)^{-1} \cdot y_1$.

Example

We will show the correctness of the obtained expressions in the following simple example.

Let's fix the four-parameter generalized Suzuki group $G = A_4(n, \theta)$ over the finite field $F_q, q = 2^5, g(x) = x^5 + x^3 + 1$. Assume that θ is the Frobenius automorphism of $F_q, \theta : \alpha \to \alpha^2$. Group operation is defined as:

$$S(a_1, a_2, a_3, a_4)S(b_1, b_2, b_3, b_4) = S(a_1 + b_1, a_2 + a_1^2 b_1 + b_2,$$
$$a_3 + a_2^2 b_1 + a_1^4 b_2 + b_3, a_4 + a_3^2 b_1 + a_2^4 b_2 + a_1^8 b_3 + b_4).$$

The inverse element is determined as:

$$S(a_1, a_2, a_3, a_4)^{-1} = S(a_1, a_2 + a_1^3, a_3 + a_2^2 a_1 + a_1^4 a_2', a_4 + a_3^2 a_1 + a_2^4 a_2' + a_1^8 a_3')$$

where $a_2' = a_2 + a_1^3, a_3' = a_3 + a_2^2 a_1 + a_1^4 a_2'$.

Let's consider the basic steps of our calculations.

Generation of public and private keys

First stage is to generate a tame logarithmic signature with the dimension of corresponding selected type $(r_{1(k)}, ..., r_{s(k)})$ and finite field F_q. The construction of arrays of logarithmic signatures is presented in [11]. For our example, we use the construction of simple logarithmic signatures without analyzing the details of their secrecy. Let's $\beta_{(k)}$

for $k = \overline{1,3}$ have the types of $(2^2, 2^3)$, $(2, 2^2, 2^2)$, $(2^2, 2, 2^2)$, $(2^2, 2^2, 2)$. They are represented as a strings and elements of the group over the field F_q in the table provided below (Table 1).

Table 1. Logarithmic signature generation

$$\beta_k = \left[B_{1(k)}, B_{2(k)}, B_{3(k)}, B_{4(k)}\right] = \left(b_{ij}\right)_{(k)}, \; \left(b_{ij}\right)_{(k)} \in A_{l=4}(n, \theta)$$

$B_{1(1)}$		$B_{1(2)}$		$B_{1(3)}$		$B_{1(4)}$	
00000	$0, 0, 0, 0$	00000	$0, 0, 0, 0$	00000	$0, 0, 0, 0$	00000	$0, 0, 0, 0$
10000	$\alpha^0, 0, 0, 0$	10000	$0, \alpha^0, 0, 0$	10000	$0, 0, \alpha^0, 0$	10000	$0, 0, 0, \alpha^0$
01000	$\alpha^1, 0, 0, 0$	01000	$0, \alpha^1, 0, 0$	$B_{2(3)}$		01000	$0, 0, 0, \alpha^1$
11000	$\alpha^{14}, 0, 0, 0$	11000	$0, \alpha^{14}, 0, 0$	00000	$0, 0, 0, 0$	11000	$0, 0, 0, \alpha^{14}$
$B_{2(1)}$		$B_{2(2)}$		11000	$0, 0, \alpha^{14}, 0$	$B_{2(4)}$	
01000	$\alpha^1, 0, 0, 0$	11000	$0, \alpha^{14}, 0, 0$	10100	$0, 0, \alpha^{28}, 0$	00000	$0, 0, 0, 0$
10100	$\alpha^{28}, 0, 0, 0$	11100	$0, \alpha^{22}, 0, 0$	01100	$0, 0, \alpha^{15}, 0$	00100	$0, 0, 0, \alpha^2$
11010	$\alpha^{26}, 0, 0, 0$	10010	$0, \alpha^5, 0, 0$	$B_{3(3)}$		$B_{3(4)}$	
00110	$\alpha^{16}, 0, 0, 0$	00110	$0, \alpha^{16}, 0, 0$	01000	$0, 0, \alpha^1, 0$	01000	$0, 0, 0, \alpha^1$
10001	$\alpha^{25}, 0, 0, 0$	$B_{3(2)}$		10010	$0, 0, \alpha^5, 0$	00110	$0, 0, 0, \alpha^{16}$
11101	$\alpha^{21}, 0, 0, 0$	10000	$0, \alpha^0, 0, 0$	01101	$0, 0, \alpha^{27}, 0$	00001	$0, 0, 0, \alpha^4$
10011	$\alpha^{18}, 0, 0, 0$	10011	$0, \alpha^{18}, 0, 0$	10111	$0, 0, \alpha^9, 0$	11011	$0, 0, 0, \alpha^{19}$
11111	$\alpha^{20}, 0, 0, 0$						

Construct random covers α_k, for the same type as $\beta_{(k)}$

$$\alpha_k = \left[A_{1(k)}, \ldots, A_{s(k)}\right] = \left(a_{ij}\right)_k = S\left(a_{ij(k)}^{(1)}, a_{ij(k)}^{(2)}, a_{ij(k)}^{(3)}, a_{ij(k)}^{(4)}\right)$$

where $a_{ij} \in A_{l=4}(n, \theta)$, $a_{ij(k)}^{(v)} \in F_q \backslash \{0\}$, $i = \overline{1, s}, j = \overline{1, r_{i(k)}}, k = \overline{1, 4}$.
In the field representation α_k has the following form (Table 2)

Table 2. Random covers construction

$$\alpha_k = \left[A_{1(k)}, \ldots, A_{s(k)}\right] = \left(a_{ij}\right)_k = S\left(a_{ij(k)}^{(1)}, a_{ij(k)}^{(2)}, a_{ij(k)}^{(3)}, a_{ij(k)}^{(4)}\right)$$

$k = 1$	$k = 2$	$k = 3$	$k = 4$
$A_{1(1)}$	$A_{1(2)}$	$A_{1(3)}$	$A_{1(4)}$
$\alpha^6, \alpha^{11}, \alpha^{17}, \alpha^{27}$	$\alpha^{17}, \alpha^5, \alpha^{26}, \alpha^{28}$	$\alpha^0, \alpha^2, \alpha^{14}, \alpha^{20}$	$\alpha^{20}, \alpha^{14}, \alpha^{30}, \alpha^{13}$
$\alpha^{11}, \alpha^5, \alpha^7, \alpha^5$	$\alpha^{20}, \alpha^{14}, \alpha^{19}, \alpha^{24}$	$\alpha^{17}, \alpha^{27}, \alpha^{16}, \alpha^{10}$	$\alpha^4, \alpha^2, \alpha^{13}, \alpha^{17}$
$\alpha^{21}, \alpha^{18}, 0, \alpha^{16}$	$\alpha^{30}, \alpha^{21}, \alpha^6, \alpha^3$	$A_{2(3)}$	$\alpha^{19}, \alpha^{13}, \alpha^{26}, \alpha^{22}$

(continued)

Table 2. (*continued*)

$$\alpha_k = [A_{1(k)}, \ldots, A_{s(k)}] = (a_{ij})_k = S\left(a_{ij(k)}^{(1)}, a_{ij(k)}^{(2)}, a_{ij(k)}^{(3)}, a_{ij(k)}^{(4)}\right)$$

$k = 1$	$k = 2$	$k = 3$	$k = 4$
$\alpha^5, \alpha^{29}, \alpha^{12}, \alpha^{16}$	$\alpha^6, \alpha^9, \alpha^{13}, \alpha^{22}$	$\alpha^{28}, \alpha^{29}, 0, \alpha^{25}$	$\alpha^6, \alpha^{28}, \alpha^{12}, \alpha^4$
$A_{2(1)}$	$A_{2(2)}$	$\alpha^{10}, \alpha^{12}, \alpha^{22}, \alpha^{30}$	$A_{2(4)}$
$\alpha^4, \alpha^7, \alpha^4, \alpha^2$	$\alpha^{30}, \alpha^{14}, \alpha^{27}, \alpha^{30}$	$\alpha^{13}, \alpha^{23}, \alpha^{19}, \alpha^{19}$	$\alpha^{18}, \alpha^1, \alpha^1, \alpha^{24}$
$\alpha^{12}, \alpha^{11}, \alpha^3, \alpha^1$	$\alpha^1, \alpha^{18}, 0, \alpha^{13}$	$\alpha^0, \alpha^{10}, \alpha^1, \alpha^{20}$	$\alpha^{26}, \alpha^{28}, \alpha^{15}, \alpha^0$
$\alpha^{18}, \alpha^{15}, \alpha^{14}, \alpha^{30}$	$\alpha^1, \alpha^{18}, \alpha^{28}, \alpha^{30}$	$A_{3(3)}$	$A_{3(4)}$
$\alpha^3, \alpha^{19}, \alpha^{26}, \alpha^2$	$\alpha^{25}, \alpha^5, \alpha^0, \alpha^{13}$	$\alpha^{11}, \alpha^{27}, \alpha^{29}, \alpha^{18}$	$\alpha^{16}, \alpha^{17}, \alpha^{29}, \alpha^{17}$
$\alpha^{11}, \alpha^{18}, \alpha^{21}, \alpha^{28}$	$A_{3(2)}$	$\alpha^5, \alpha^1, \alpha^{12}, \alpha^{22}$	$\alpha^{18}, \alpha^0, \alpha^1, \alpha^{15}$
$\alpha^{16}, \alpha^{18}, \alpha^{10}, \alpha^{24}$	$\alpha^3, \alpha^{29}, \alpha^{25}, 0$	$\alpha^{30}, \alpha^{18}, \alpha^6, \alpha^{11}$	$\alpha^4, \alpha^9, \alpha^{23}, \alpha^{19}$
$\alpha^{17}, \alpha^{16}, 0, \alpha^{27}$	$\alpha^{25}, \alpha^{19}, \alpha^{23}, \alpha^2$	$0, 0, \alpha^{17}, \alpha^{23}$	$\alpha^{19}, \alpha^{20}, \alpha^{30}, \alpha^{10}$
$\alpha^{25}, \alpha^{17}, \alpha^8, \alpha^{12}$			

Choose random $A_l(n, \theta)$ $t_{0(k)}, t_{1(k)}, \ldots, t_{s(k)} \in A_l(n, \theta)$, $s_{(k)}$, $k = \overline{1,4}$ and $t_{2(1)} = t_{0(2)}$, $t_{3(2)} = t_{0(3)}$, $t_{3(3)} = t_{0(4)}$ (Table 3)

Table 3. Random t vectors

$t_{0(k)}, t_{1(k)}, \ldots, t_{s(k)} \in A_{l=4}(n, \theta)$, $s_{(k)}$, $k = \overline{1,4}$			
$k = 1$	$k = 2$	$k = 3$	$k = 4$
$\alpha^1, \alpha^5, \alpha^{17}, \alpha^{16}$	$\alpha^{13}, \alpha^0, \alpha^{28}, \alpha^{10}$	$\alpha^9, \alpha^4, \alpha^9, \alpha^{20}$	$\alpha^{12}, \alpha^{15}, \alpha^{17}, \alpha^6$
$\alpha^{25}, \alpha^{17}, \alpha^{23}, \alpha^{27}$	$\alpha^{30}, \alpha^2, \alpha^{17}, \alpha^2$	$\alpha^{14}, \alpha^{28}, \alpha^{17}, \alpha^{22}$	$\alpha^{22}, \alpha^{30}, \alpha^{22}, \alpha^{16}$
$\alpha^{13}, \alpha^0, \alpha^{28}, \alpha^{10}$	$\alpha^6, \alpha^7, \alpha^{30}, \alpha^{18}$	$\alpha^{26}, \alpha^5, \alpha^{16}, \alpha^{30}$	$\alpha^{24}, \alpha^{29}, \alpha^{15}, \alpha^{30}$
	$\alpha^9, \alpha^4, \alpha^9, \alpha^{20}$	$\alpha^{12}, \alpha^{15}, \alpha^{17}, \alpha^6$	$\alpha^3, 0, \alpha^{14}, \alpha^9$

The inverse elements $t_{0(k)}^{-(k)}, t_{1(k)}^{-(k)}, \ldots, t_{s(k)}^{-(k)}$ of the group $A_4(n, \theta)$ were computed with reference below (Table 4):

Table 4. Computing of inverse elements $t_{0(k)}^{-(k)}, t_{1(k)}^{-(k)}, \ldots, t_{s(k)}^{-(k)}$

$\tau_{0(k)}^{-(k)}, \tau_{1(k)}^{-(k)}, \ldots, \tau_{s(k)}^{-(k)}$			
$k = 1$	$k = 2$	$k = 3$	$k = 4$
$\alpha^1, \alpha^0, \alpha^{22}, \alpha^{21}$	$\alpha^{13}, \alpha^0, \alpha^7, \alpha^{24}$	$\alpha^9, \alpha^4, \alpha^9, \alpha^{25}$	$\alpha^{12}, \alpha^{15}, \alpha^{17}, \alpha^6$
$\alpha^{25}, \alpha^7, \alpha^3, \alpha^{15}$	$\alpha^{30}, \alpha^2, \alpha^{15}, \alpha^{21}$	$\alpha^{14}, \alpha^{28}, \alpha^{17}, \alpha^{21}$	$\alpha^{22}, \alpha^{30}, \alpha^{22}, \alpha^{16}$
$\alpha^{13}, \alpha^{19}, \alpha^7, \alpha^{24}$	$\alpha^6, \alpha^7, \alpha^{28}, \alpha^{24}$	$\alpha^{26}, \alpha^5, \alpha^{16}, \alpha^{13}$	$\alpha^{24}, \alpha^{29}, \alpha^{15}, \alpha^{30}$
	$\alpha^9, \alpha^4, \alpha^8, \alpha^{25}$	$\alpha^{12}, \alpha^{15}, \alpha^{17}, \alpha^{30}$	$\alpha^3, 0, \alpha^{14}, \alpha^9$

Similarly, we choose random $\tau_{0(k)}, \tau_{1(k)}, \ldots, \tau_{s(k)} \in A_l(n, \theta)$, $s_{(k)}$, $k = \overline{1,4}$ and $t_{2(1)} = t_{0(2)}, t_{3(2)} = t_{0(3)}, t_{3(3)} = t_{0(4)}$:

and the inverse elements $\tau_{0(k)}^{-(k)}, \tau_{1(k)}^{-(k)}, \ldots, \tau_{s(k)}^{-(k)}$ (Table 5):

Table 5. Computing of random τ vectors $\tau_{0(k)}, \tau_{1(k)}, \ldots, \tau_{s(k)} \in A(P_\infty) \backslash Z$

$\tau_{0(k)}, \tau_{1(k)}, \ldots, \tau_{s(k)} \in A(P_\infty) \backslash Z$, $s_{(k)}$, $k = \overline{1,4}$			
$k = 1$	$k = 2$	$k = 3$	$k = 4$
$\alpha^4, \alpha^{22}, \alpha^7, \alpha^{12}$	$\alpha^{29}, \alpha^{21}, \alpha^{30}, \alpha^{13}$	$\alpha^2, \alpha^{17}, \alpha^{22}, \alpha^2$	$\alpha^{20}, 0, \alpha^3, \alpha^0$
$\alpha^8, 0, \alpha^{13}, \alpha^{16}$	$\alpha^{24}, \alpha^{20}, \alpha^{17}, \alpha^{25}$	$0, \alpha 22, \alpha^{16}, \alpha^{24}$	$\alpha^{21}, \alpha^{16}, \alpha^{12}, \alpha^{16}$
$\alpha^{29}, \alpha^{21}, \alpha^{30}, \alpha^{13}$	$\alpha^4, \alpha^7, \alpha^{16}, \alpha^{30}$	$\alpha^6, \alpha^{21}, \alpha^{25}, \alpha^{18}$	$\alpha^{16}, \alpha^{28}, \alpha^{19}, \alpha^{16}$
	$\alpha^2, \alpha^{17}, \alpha^{22}, \alpha^2$	$\alpha^{20}, 0, \alpha^3, \alpha^0$	$\alpha^{28}, \alpha^{17}, \alpha^{26}, \alpha^4$

Table 6. Computing of inverse elements $\tau_{0(k)}^{-(k)}, \tau_{1(k)}^{-(k)}, \ldots, \tau_{s(k)}^{-(k)}$

$\tau_{0(k)}^{-(k)}, \tau_{1(k)}^{-(k)}, \ldots, \tau_{s(k)}^{-(k)}$			
$k = 1$	$k = 2$	$k = 3$	$k = 4$
$\alpha^4, \alpha^{18}, \alpha^9, \alpha^0$	$\alpha^{29}, \alpha^{21}, \alpha^2, \alpha^5$	$\alpha^2, \alpha^{17}, \alpha^{22}, \alpha^{11}$	$\alpha^{20}, 0, \alpha^3, \alpha^0$
$\alpha^8, \alpha^{24}, \alpha^2, \alpha^{30}$	$\alpha^{24}, \alpha^{20}, \alpha^{22}, \alpha^{29}$	$0, \alpha^{22}, \alpha^{16}, \alpha^2$	$\alpha^{21}, \alpha^{16}, \alpha^{12}, \alpha^{16}$
$\alpha^{29}, \alpha^{15}, \alpha^2, \alpha^5$	$\alpha^4, \alpha^7, \alpha^{12}, \alpha^{28}$	$\alpha^6, \alpha^{21}, \alpha^{25}, \alpha^3$	$\alpha^{16}, \alpha^{28}, \alpha^{19}, \alpha^{16}$
	$\alpha^2, \alpha^{17}, \alpha^{24}, \alpha^{11}$	$\alpha^{20}, 0, \alpha^3, \alpha^{22}$	$\alpha^{28}, \alpha^{17}, \alpha^{26}, \alpha^4$

Construct random covers w_k, for the same type as $\beta_{(k)}$

$$w_{(k)} = \left[W_{1(k)}, \ldots, W_{s(k)}\right] = \left(w_{ij}\right)_{(k)} = S\left(w_{ij(k)}^{(1)}, w_{ij(k)}^{(2)}, \ldots, w_{ij(k)}^{(l)}\right), \text{ where } w_{ij} \in$$
$A_{l=4}(n, \theta)$, $w_{ij(k)}^{(v)} \in F_q$, $i = \overline{0, s(k)}$, $j = \overline{1, r_{i(k)}}$, $k = \overline{1,4}$ (Table 6 and 7).

Table 7. Construct random covers w_k

$w_{(k)} = \left[W_{1(k)}, \ldots, W_{s(k)}\right] = \left(w_{ij}\right)_{(k)} = S\left(w_{ij(k)}^{(1)}, \ldots, w_{ij(k)}^{(4)}\right)$			
$k = 1$	$k = 2$	$k = 3$	$k = 4$
$W_{1(1)}$	$W_{1(2)}$	$W_{1(3)}$	$W_{1(4)}$
$\alpha^{20}, \alpha^{20}, \alpha^{12}, \alpha^4$	$\alpha^9, \alpha^{28}, \alpha^{27}, \alpha^2$	$\alpha^3, \alpha^2, \alpha^{10}, 0$	$\alpha^{30}, \alpha^{14}, \alpha^1, \alpha^{28}$
$\alpha^7, \alpha^9, \alpha^{17}, \alpha^{20}$	$\alpha^{16}, \alpha^{13}, \alpha^6, \alpha^{21}$	$\alpha^5, \alpha^{10}, \alpha^{19}, \alpha^{16}$	$\alpha^6, \alpha^{28}, \alpha^{30}, \alpha^{20}$
$\alpha^{25}, \alpha^6, \alpha^{23}, \alpha^{27}$	$\alpha^{25}, 0, \alpha^4, \alpha^{27}$	$W_{2(3)}$	$\alpha^{13}, \alpha^{19}, \alpha^{26}, \alpha^{11}$
$\alpha^3, \alpha^0, \alpha^{23}, \alpha^{29}$	$\alpha^1, \alpha^0, \alpha^{17}, \alpha^{17}$	$\alpha^{12}, \alpha^{20}, \alpha^{14}, \alpha^3$	$\alpha^{16}, \alpha^{27}, \alpha^9, \alpha^{21}$

(continued)

Table 7. (*continued*)

$$w_{(k)} = \left[W_{1(k)}, \ldots, W_{s(k)}\right] = \left(w_{ij}\right)_{(k)} = S\left(w_{ij(k)}^{(1)}, \ldots, w_{ij(k)}^{(4)}\right)$$

$W_{2(1)}$	$W_{2(2)}$	$\alpha^{23}, \alpha^{12}, \alpha^5, \alpha^{27}$	$W_{2(4)}$
$\alpha^7, \alpha^{21}, \alpha^6, \alpha^{21}$	$\alpha^{21}, \alpha^{14}, \alpha^{14}, \alpha^0$	$\alpha^2, \alpha^3, \alpha^{24}, \alpha^{16}$	$\alpha^2, \alpha^{21}, \alpha^8, \alpha^{29}$
$\alpha^{18}, \alpha^{21}, \alpha^{22}, \alpha^6$	$\alpha^{19}, \alpha^{29}, \alpha^{19}, \alpha^{13}$	$\alpha^{12}, \alpha^5, \alpha^{21}, \alpha^{14}$	$\alpha^4, \alpha^2, \alpha^1, \alpha^{23}$
$\alpha^{18}, \alpha^{19}, \alpha^{12}, \alpha^{15}$	$\alpha^{25}, \alpha^{26}, \alpha^{12}, \alpha^{17}$	$W_{3(3)}$	$W_{3(4)}$
$\alpha^{16}, \alpha^{12}, \alpha^{14}, \alpha^6$	$\alpha^{10}, \alpha^{19}, \alpha^{23}, \alpha^4$	$\alpha^{14}, \alpha^6, \alpha^0, \alpha^{17}$	$0, \alpha^0, \alpha^{25}, \alpha^3$
$\alpha^{23}, \alpha^4, \alpha^1, \alpha^{30}$	$W_{3(2)}$	$\alpha^{17}, \alpha^{13}, \alpha^7, \alpha^4$	$\alpha^3, \alpha^{19}, \alpha^{17}, \alpha^{24}$
$\alpha^5, \alpha^{26}, \alpha^6, \alpha^{19}$	$\alpha^{28}, \alpha^0, \alpha^{13}, \alpha^{17}$	$\alpha^{25}, \alpha^{24}, \alpha^{27}, \alpha^8$	$\alpha^{28}, \alpha^{28}, \alpha^{14}, \alpha^{26}$
$\alpha^{22}, \alpha^{17}, \alpha^{13}, \alpha^{21}$	$\alpha^{14}, \alpha^0, \alpha^3, \alpha^3$	$\alpha^{13}, 0, \alpha^{21}, \alpha^7$	$\alpha^{24}, \alpha^{18}, \alpha^{27}, \alpha^{13}$
$\alpha^{28}, \alpha^{27}, \alpha^9, \alpha^{24}$			

The next step is to calculate the arrays h_k. Within the condition of the example, we obtain:

$$h_{(k)} = \left[h_{1(k)}, \ldots, h_{s(k)}\right] = t_{(i-1)(k)}^{-(k)} \circ^{(k)} \left(w_{ij}\right)_{(k)} \circ^{(k)} \left(b_{ij}\right)_{(k)} \circ^{(k)} t_{i(k)}$$

$i = \overline{1, s(k)}, j = \overline{1, r_{i(k)}}, k = \overline{1,4}.$

Let's a homomorphic cryptographic transformation for a field element $e \Rightarrow \rho_i e$ where ρ_i is a secret parameter. The transformation is chosen to be the simplest. You can also use more complex homomorphic transformations with respect to the addition operation. We define homomorphic cryptographic transformation for a group element S as

$$f(S(e_1, e_2, e_3, e_4)) = S(\rho_1 e_1, \rho_2 e_2, \rho_3 e_3, \rho_4 e_4),$$

and let's $\rho = (\rho_1, \rho_2, \rho_3, \rho_4) = \left(\alpha^4, \alpha^5, \alpha^6, \alpha^7\right)$.

Let's a homomorphic cryptographic transformation for a field element $e \Rightarrow \rho_i e$ where ρ_i is a secret parameter. The transformation is chosen to be the simplest (Table 8).

You can also use more complex homomorphic transformations with respect to the addition operation. We define homomorphic cryptographic transformation for a group element S as

$$f(S(e_1, e_2, e_3, e_4)) = S(\rho_1 e_1, \rho_2 e_2, \rho_3 e_3, \rho_4 e_4),$$

and let's $\rho = (\rho_1, \rho_2, \rho_3, \rho_4) = \left(\alpha^4, \alpha^5, \alpha^6, \alpha^7\right)$.

Next, we compute the arrays g_k via the homomorphic transformation

$$g_{(k)} = \left[g_{1(k)}, \ldots, g_{s(k)}\right] = \tau_{(i-1)(k)}^{-(k)} \circ^{(k)} f\left(w_{ij}\right)_{(k)} \circ^{(k)} \tau_{i(k)}$$

$i = \overline{1, s(k)}, j = \overline{1, r_{i(k)}}, k = \overline{1,4}.$ See the Table 9 for the results.

An output public key (a_k, h_k, g_k), and a private key $\left[f, \beta_{(k)}, \left(t_{0(k)}, \ldots, t_{s(k)}\right), \left(\tau_{0(k)}, \ldots, \tau_{s(k)}\right)\right], k = \overline{1,4}.$

Table 8. Construct arrays h_k

$h_k = S(h_{ij(k)}^{(1)}, h_{ij(k)}^{(2)}, h_{ij(k)}^{(3)}, h_{ij(k)}^{(4)})$

$k = 1$	$k = 2$	$k = 3$	$k = 4$
$h_{1(1)}$	**$h_{1(2)}$**	**$h_{1(3)}$**	**$h_{1(4)}$**
$\alpha^{16}, \alpha^{20}, \alpha^{22}, \alpha^{30}$	$\alpha^{24}, 0, \alpha^{16}, 0$	$\alpha^{27}, \alpha^{25}, \alpha^{27}, \alpha^{30}$	$\alpha^{7}, \alpha^{25}, \alpha^{9}, \alpha^{19}$
$\alpha^{20}, \alpha^{7}, \alpha^{21}, \alpha^{15}$	$\alpha^{7}, \alpha^{25}, \alpha^{21}, \alpha^{3}$	$\alpha^{21}, \alpha^{15}, \alpha^{20}, \alpha^{14}$	$\alpha^{26}, \alpha^{21}, \alpha^{26}, 0$
$0, \alpha^{27}, \alpha^{26}, \alpha^{13}$	$\alpha^{4}, \alpha^{22}, 0, \alpha^{21}$	**$h_{2(3)}$**	$\alpha^{16}, \alpha^{5}, \alpha^{30}, \alpha^{10}$
$\alpha^{17}, \alpha^{16}, \alpha^{28}, \alpha^{26}$	$\alpha^{14}, \alpha^{22}, \alpha^{3}, \alpha^{5}$	$\alpha^{27}, \alpha^{10}, \alpha^{21}, \alpha^{23}$	$\alpha^{13}, \alpha^{2}, \alpha^{1}, \alpha^{29}$
$h_{2(1)}$	**$h_{2(2)}$**	$\alpha^{15}, \alpha^{6}, \alpha^{12}, \alpha^{9}$	**$h_{2(4)}$**
$\alpha^{26}, 0, \alpha^{29}, \alpha^{11}$	$\alpha^{25}, \alpha^{5}, \alpha^{3}, \alpha^{26}$	$\alpha^{16}, \alpha^{2}, \alpha^{7}, \alpha^{17}$	$\alpha^{20}, \alpha^{5}, \alpha^{19}, \alpha^{6}$
$\alpha^{17}, \alpha^{7}, \alpha^{26}, \alpha^{29}$	$\alpha^{9}, \alpha^{2}, \alpha^{12}, \alpha^{14}$	$\alpha^{27}, \alpha^{28}, \alpha^{28}, \alpha^{11}$	$\alpha^{26}, \alpha^{8}, \alpha^{14}, \alpha^{6}$
$\alpha^{27}, \alpha^{11}, \alpha^{28}, \alpha^{16}$	$\alpha^{21}, \alpha^{26}, \alpha^{25}, \alpha^{21}$	**$h_{3(3)}$**	**$h_{3(4)}$**
$\alpha^{2}, \alpha^{3}, \alpha^{11}, \alpha^{4}$	$\alpha^{13}, \alpha^{12}, \alpha^{22}, \alpha^{7}$	$\alpha^{27}, \alpha^{9}, \alpha^{21}, \alpha^{15}$	$\alpha^{30}, \alpha^{26}, \alpha^{30}, \alpha^{14}$
$\alpha^{19}, \alpha^{16}, \alpha^{25}, \alpha^{5}$	**$h_{3(2)}$**	$\alpha^{7}, \alpha^{8}, \alpha^{4}, \alpha^{4}$	$\alpha^{24}, \alpha^{25}, \alpha^{9}, \alpha^{18}$
$\alpha^{8}, \alpha^{8}, \alpha^{19}, \alpha^{19}$	$\alpha^{29}, \alpha^{9}, \alpha^{1}, \alpha^{12}$	$\alpha^{2}, \alpha^{10}, \alpha^{30}, \alpha^{24}$	$\alpha^{25}, \alpha^{11}, \alpha^{15}, \alpha^{6}$
$\alpha^{8}, \alpha^{10}, \alpha^{1}, \alpha^{30}$	$\alpha^{16}, \alpha^{28}, \alpha^{1}, \alpha^{3}$	$0, \alpha^{11}, \alpha^{12}, \alpha^{21}$	$\alpha^{3}, \alpha^{10}, \alpha^{10}, \alpha^{22}$
$\alpha^{12}, \alpha^{27}, \alpha^{0}, \alpha^{21}$			

Table 9. Construct arrays g_k

$g_k = S(g_{ij(k)}^{(1)}, g_{ij(k)}^{(2)}, g_{ij(k)}^{(3)}, g_{ij(k)}^{(4)})$

$k = 1$	$k = 2$	$k = 3$	$k = 4$
$g_{1(1)}$	**$g_{1(2)}$**	**$g_{1(3)}$**	**$g_{1(4)}$**
$\alpha^{27}, \alpha^{21}, \alpha^{17}, \alpha^{13}$	$\alpha^{14}, \alpha^{16}, \alpha^{7}, \alpha^{18}$	$\alpha^{5}, \alpha^{6}, \alpha^{22}, \alpha^{30}$	$0, \alpha^{21}, \alpha^{19}, \alpha^{9}$
$\alpha^{28}, \alpha^{18}, \alpha^{2}, \alpha^{1}$	$\alpha^{5}, \alpha^{25}, \alpha^{18}, 0$	$\alpha^{18}, \alpha^{18}, \alpha^{8}, \alpha^{7}$	$\alpha^{19}, \alpha^{3}, \alpha^{20}, \alpha^{19}$
$0, \alpha^{17}, \alpha^{1}, \alpha^{13}$	$\alpha^{24}, \alpha^{3}, \alpha^{1}, \alpha^{13}$	**$g_{2(3)}$**	$\alpha^{4}, \alpha^{4}, \alpha^{30}, \alpha^{30}$
$\alpha^{22}, \alpha^{9}, \alpha^{29}, \alpha^{26}$	$\alpha^{20}, \alpha^{0}, 0, \alpha^{23}$	$\alpha^{12}, \alpha^{0}, \alpha^{1}, \alpha^{0}$	$\alpha^{21}, \alpha^{23}, \alpha^{4}, \alpha^{3}$
$g_{2(1)}$	**$g_{2(2)}$**	$\alpha^{2}, \alpha^{3}, \alpha^{6}, 0$	**$g_{2(4)}$**
$\alpha^{20}, \alpha^{29}, \alpha^{17}, \alpha^{13}$	$\alpha^{9}, \alpha^{5}, \alpha^{25}, \alpha^{30}$	$0, \alpha^{29}, \alpha^{5}, \alpha^{11}$	$\alpha^{5}, \alpha^{1}, \alpha^{15}, \alpha^{5}$
$\alpha^{21}, \alpha^{0}, \alpha^{25}, \alpha^{28}$	$\alpha^{1}, \alpha^{8}, \alpha^{7}, \alpha^{17}$	$\alpha^{12}, \alpha^{14}, \alpha^{26}, \alpha^{23}$	$\alpha^{0}, \alpha^{2}, \alpha^{3}, \alpha^{30}$
$\alpha^{21}, \alpha^{27}, \alpha^{21}, \alpha^{21}$	$\alpha^{15}, \alpha^{10}, \alpha^{13}, \alpha^{9}$	**$g_{3(3)}$**	**$g_{3(4)}$**
$\alpha^{11}, \alpha^{30}, \alpha^{22}, \alpha^{5}$	$\alpha^{11}, \alpha^{23}, \alpha^{29}, \alpha^{18}$	$\alpha^{30}, \alpha^{17}, \alpha^{26}, \alpha^{2}$	$\alpha^{5}, \alpha^{30}, \alpha^{25}, \alpha^{11}$
$\alpha^{15}, \alpha^{24}, \alpha^{17}, \alpha^{24}$	**$g_{3(2)}$**	$\alpha^{8}, \alpha^{23}, \alpha^{16}, \alpha^{9}$	$\alpha^{2}, \alpha^{0}, \alpha^{12}, \alpha^{9}$
$\alpha^{7}, \alpha^{30}, \alpha^{20}, \alpha^{24}$	$\alpha^{27}, \alpha^{24}, \alpha^{6}, \alpha^{9}$	$\alpha^{22}, \alpha^{9}, \alpha^{9}, \alpha^{10}$	$\alpha^{26}, \alpha^{18}, \alpha^{11}, \alpha^{17}$
$\alpha^{19}, \alpha^{19}, \alpha^{3}, \alpha^{2}$	$\alpha^{7}, \alpha^{24}, \alpha^{25}, \alpha^{26}$	$\alpha^{13}, \alpha^{21}, \alpha^{11}, \alpha^{26}$	$\alpha^{16}, \alpha^{10}, \alpha^{30}, \alpha^{14}$
$\alpha^{6}, \alpha^{10}, \alpha^{17}, \alpha^{17}$			

Encryption

Input: a message $m \in A_l(n, \theta)$, $m = S(m_1, m_2, m_3, m_4)$, $m_i \in F_q$ and the public key $[f_k, (a_k, h_k, g_k)]$, $k = \overline{1, 4}$.

Let $m = (\alpha^1, \alpha^2, \alpha^3, \alpha^4) = S(\alpha^1, \alpha^2, \alpha^3, \alpha^4)$.

Choose a random $R = (R_1, R_2, R_3, R_4) = (10, 20, 30, 14)$.

We obtain the following R_i expansions for given types of $(r_{1(k)}, ..., r_{s(k)})$, $k = \overline{1, 4}$

$$R_1 = \left(R_{1(1)}, R_{2(1)}\right) = (2, 2) = 10,$$

$$R_2 = \left(R_{1(2)}, R_{2(2)}, R_{3(2)}\right) = (0, 1, 1) = 20,$$

$$R_3 = \left(R_{1(3)}, R_{2(3)}, R_{3(3)}\right) = (0, 3, 3) = 30.$$

$$R_4 = \left(R_{1(4)}, R_{2(4)}, R_{3(4)}\right) = (2, 1, 1) = 14$$

Compute the cipher text:

$$y_1 = a'(R) \cdot m = a_1'(R_1) \cdot a_2'(R_2) \cdot a_3'(R_3) \cdot a_4'(R_4) \cdot m =$$
$$S\left(\alpha^7, \alpha^6, \alpha^{22}, \alpha^{11}\right)$$

where:

$$a_1'(R_1) = a_1(10) = a_{1(1)}(2)a_{2(1)}(2) = S\left(\alpha^{23}, \alpha^{13}, \alpha^{20}, \alpha^{20}\right),$$

$$a_2'(R_2) = a_2(20) = a_{1(2)}(0)a_{2(2)}(1)a_{3(2)}(1) = S\left(\alpha^{26}, \alpha^3, \alpha^5, \alpha^{29}\right),$$

$$a_3'(R_3) = a_3(30) = a_{1(3)}(0)a_{2(3)}(3)a_{3(3)}(3) = S\left(0, \alpha^{27}, \alpha^8, \alpha^4\right),$$

$$a_4'(R_4) = a_4(14) = a_{1(4)}(2)a_{2(4)}(1)a_{3(4)}(1) = S\left(\alpha^5, \alpha^{12}, \alpha^{21}, \alpha^{16}\right).$$

Calculate

$$y_2 = h_1(R_1) \circ^{(1)} \left(h_2(R_2) \circ^{(2)} \left(h_3(R_3) \circ^{(3)} h_4(R_4)\right)\right) = S\left(0, \alpha^8, \alpha^{16}, \alpha^{17}\right)$$

The components $h_k'(R_k)$ are calculated similarly to $a_k'(R_k)$ components, but using the appropriate multiplication operation. Compute the component y_3:

$$y_3 = g_1(R_1) \circ^{(1)} \left(g_2(R_2) \circ^{(2)} \left(g_3(R_3) \circ^{(3)} g_4(R_4)\right)\right) = S\left(\alpha^{16}, \alpha^{14}, \alpha^1, \alpha^4\right).$$

We obtained output $y_1 = (\alpha^7, \alpha^6, \alpha^{22}, \alpha^{11})$, $y_2 = (0, \alpha^8, \alpha^{16}, \alpha^{17})$, $y_3 = (\alpha^{16}, \alpha^{14}, \alpha^1, \alpha^4)$.

Decryption

Input: a ciphertext (y_1, y_2, y_3) and private key $[f, \beta_{(k)}, t_{i(k)}, \tau_{i(k)}]$, $i = \overline{0, s(k)}$, $k = \overline{1, 4}$.

Output: the message $m \in A(P_\infty)$ corresponding to ciphertext (y_1, y_2, y_3).
To decrypt a message m, we need to restore random numbers $R = (R_1, R_2, R_3)$.
Compute

$$D^{(1)}(R) = t_{0(1)} \circ^{(1)} y_2 \circ^{(4)} t_{s(4)}^{-(4)} = S(\alpha^{29}, \alpha^8, \alpha^{24}, \alpha^{28}),$$

$$G^{(1)}(R) = \tau_{0(1)} \circ^{(1)} y_3 \circ^{(4)} \tau_{s(4)}^{-(4)} = S(\alpha^{18}, \alpha^5, \alpha^7, \alpha^{30}),$$

$$D^{(1)}(R)' = D^{(1)}(R) \circ^{(1)} \hat{f}(G^{(1)}(R))^{-(1)} = S(\alpha^5, \alpha^{22}, \alpha^{21}, \alpha^0).$$

Restore R_1 with $\beta_{(1)}(R_1) = \sum\limits_{i=1, j=R_{i(1)}}^{s(1)} \beta_{ij(1)}$ using $\beta_{(1)}(R_1)^{-1}$, because β_1 is simple.

We get $\beta_1(R_1) = \alpha^5 = (10010)$. Perform inverse calculations $\beta_{(1)}(R_1)^{-1}$.
10|010 $R_1 = (*, 2)$
11|010 row 1 from $B_{4(1)}$
10|010−11|010 = 01|000 $R_1 = (2, 2)$
We get $\beta_1(R_1)^{-1} = (2, 2) = 10$
For further calculation, it is necessary to remove the component $h_1'(R_1)$ from y_2 and
$g_1'(R_1)$ from y_3.
Compute

$$y_2^{(1)} = h_1(R_1)^{-(1)\circ} \cdot y_2^\circ = S(\alpha^{26}, \alpha^{16}, \alpha^{17}, \alpha^{19}),$$

$$y_3^{(1)} = g_1(R_1)^{-(1)\circ} \cdot y_3^\circ = S(\alpha^{19}, \alpha^{18}, \alpha^{12}, \alpha^{19}),$$

$$D^{(2)}(R) = t_{0(2)} \circ^{(2)} y_2^{(1)} \circ^{(4)} t_{s(4)}^{-(4)} = S(\alpha^{26}, \alpha^{18}, \alpha^{16}, \alpha^2),$$

$$G^{(2)}(R) = \tau_{0(2)} \circ^{(2)} y_3^{(1)} \circ^{(4)} \tau_{s(4)}^{-(4)} = S(\alpha^{30}, \alpha^{27}, \alpha^0, \alpha^{11}),$$

$$D^{(2)}(R)' = D^{(2)}(R) \circ^{(2)} \hat{f}(G^{(2)}(R))^{-(2)} = S(0, \alpha^{12}, \alpha^4, \alpha^{30})$$

and restore R_2 with $\beta_{(2)}(R_2) = \sum\limits_{i=1, j=R_{i(2)}}^{s(2)} \beta_{ij(2)}$ using $\beta_{(2)}(R_2)^{-1}$, because β_2 is

simple. We get $\beta_2(R_2) = \alpha^{12} = (01111)$. Restore R_2 with $\beta_2(R_2)$. We use the same
calculations as in the example for $\beta_2(R_2)^{-1}$, and we get:
01|11|1 $R_2 = (*, *, 1)$
10|01|1 row 1 from $B_{3(2)}$
01|11|1−10|01|1 = 11|10|0 $R_2 = (*, 1, 1)$
11|10|0 row 0 from $B_{3(2)}$
11|10|0−11|10|0 = 00|00|0 $R_2 = (0, 1, 1)$
We get $\beta_2(R_2)^{-1} = (0, 1, 1) = 20$.
Remove the component $h_2'(R_2)$ from $y_2^{(1)}$ and $g_2'(R_2)$ from $y_3^{(1)}$. We get

$$y_2^{(2)} = h_3(R_3)^{-(2)\circ} \cdot y_2^{(1)\circ} = S(\alpha^{19}, \alpha^{18}, \alpha^{22}, \alpha^{15}),$$

$$y_3^{(2)} = g_3(R_3)^{-(2)\circ} \cdot y_3^{(1)\circ} = S(\alpha^{21}, \alpha^{10}, \alpha^0, \alpha^{19}),$$

$$D^{(3)}(R) = t_{0(3)} \circ^{(3)} y_2^{(2)} \circ^{(4)} t_{s(4)}^{-(4)} = S(\alpha^{23}, \alpha^5, \alpha^{18}, \alpha^{21}),$$

$$G^{(3)}(R) = \tau_{0(3)} \circ^{(3)} y_3^{(2)} \circ^{(4)} \tau_{s(4)}^{-(4)} = S(\alpha^{21}, \alpha^{10}, \alpha^7, \alpha^{13}),$$

$$D^{(3)}(R)' = D^{(3)}(R) \circ^{(3)} \hat{f}(G^{(3)}(R))^{-(3)} = S(0, 0, \alpha^{19}, \alpha^6)$$

We get $\beta_3(R_3) = \alpha^{19} = (11011)$.
Perform inverse calculations $\beta_3(R_3)^{-1}$.
$1|10|11\ R_3 = (*, *, 3)$
$1|01|11$ row 3 from $B_{3(3)}$
$1|10|11 - 1|01|11 = 0|11|00\ R_3 = *, 3, 3)$
$0|11|00$ row 3 from $B_{2(3)}$
$0|11|00 - 0|11|00 = 0|00|00\ R_3 = (0, 3, 3)$
We get $\beta_3(R_3)^{-1} = (0, 3, 3) = 30$.
Remove the component $h_3'(R_3)$ from $y_2^{(2)}$ and $g_3'(R_3)$ from $y_3^{(2)}$.
As a result, we get:

$$y_2^{(3)} = h_3(R_3)^{-(3)\circ} \cdot y_2^{(2)\circ} = S(\alpha^{19}, \alpha^1, \alpha^{29}, \alpha^{17}),$$

$$y_3^{(3)} = g_3(R_3)^{-(3)\circ} \cdot y_3^{(2)\circ} = S(\alpha^{13}, \alpha^{13}, \alpha^0, \alpha^{16}),$$

$$D^{(4)}(R) = t_{0(4)} \circ^{(4)} y_2^{(3)} \circ^{(4)} t_{s(4)}^{-(4)} = S(\alpha^7, \alpha^2, \alpha^{25}, \alpha^{21}),$$

$$G^{(4)}(R) = \tau_{0(4)} \circ^{(3)} y_3^{(3)} \circ^{(4)} \tau_{s(4)}^{-(4)} = S(\alpha^{11}, \alpha^7, \alpha^0, \alpha^{16}),$$

$$D^{(3)}(R)' = D^{(4)}(R) \circ^{(4)} \hat{f}(G^{(4)}(R))^{-(4)} = S(0, 0, 0, \alpha^{29})$$

01010
We get $\beta_4(R_4) = \alpha^{29} = (01010)$. Perform inverse calculations $\beta_4(R_4)^{-1}$.
$01|0|10\ R_3 = (*, *, 1)$
$00|1|10$ row 1 from $B_{3(4)}$
$01|0|10 - 00|1|10 = 01|1|00\ R_3 = (*, 1, 1)$
$00|1|00$ row 1 from $B_{2(4)}$
$01|1|00 - 00|1|00 = 01|0|00\ R_3 = (2, 1, 1)$
We get $\beta_4(R_4)^{-1} = (2, 1, 1) = 14$.
Receive a message $m = a'(R)^{-1}y_1 = S(\alpha^1, \alpha^2, \alpha^3, \alpha^4)$.

3 Security Parameters Analysis and Cost Estimation

Consider a brute force attack of key recovery. There are three possible schemes for such an attack.

Brute force attack on cipher text. By selecting $R = (R_1, R_2, ..., R_l)$ try to decipher the text $y'_1 = \alpha'(R') \cdot m = \alpha'_1(R'_1) \cdot \alpha'_2(R'_2) ... \alpha'_l(R'_l) \cdot m$. The covers $\alpha_k = (a_{ij})_k = S\left(a_{ij(k)}^{(1)}, a_{ij(k)}^{(2)}, ..., a_{ij(k)}^{(l)}\right)$ are selected randomly and the value is determined by multiplication in a group with no coordinate constraints. The resulting vector $\alpha'(R')$ depends on all components $\alpha'_i(R'_i)$. Enumeration of key values $R = (R_1, R_2, ..., R_l)$ has an estimation of complexity. For a practical attack, the message m is also unknown and has uncertainty to choose from q^l. This makes a brute-force attack on a key infeasible. If we take an attack model with a known text, then the attack complexity still remains the same and equal to q^l.

Brute force attack on the cyphertext y_2. Select $R = (R_1, R_2, ..., R_l)$ to match y_2. The vector y_2 has a following definition over the components $\alpha'_i(R_i)$

$$
y_2 = S\left(\sum_{k=1}^{l} \sum_{i=1, j=R_{i(k)}}^{s(k)} w_{ij(k)}^{(1)} + \sum_{i=1, j=R_{i(1)}}^{s(1)} \beta_{ij(1)}, \sum_{k=1}^{l} \sum_{i=1, j=R_{i(k)}}^{s(k)} w_{ij(k)}^{(2)} + \right.
$$

$$
\left. \sum_{i=1, j=R_{i(2)}}^{s(2)} \beta_{ij(2)} + *, \dots, \sum_{k=1}^{l} \sum_{i=1, j=R_{i(k)}}^{s(k)} w_{ij(k)}^{(l)} + \sum_{i=1, j=R_{i(l)}}^{s(l)} \beta_{ij(l)} + * \right)
$$

The values of the coordinates y_2 are defined by calculations over the vectors $w'_1(R_1), w'_2(R_2), ..., w'_l(R_l)$. The keys $R = (R_1, R_2, ..., R_l)$ are bound and changes in any of them leads to change y_2. The brute force attack on key R has a complexity equal to q^l.

Brute force attack on the ciphertext y_3. Select $R = (R_1, R_2, ..., R_l)$ to match y_3. The vector y_3 has a following definition over the components $\rho_i w'_i(R_i)$

$$
y_3 = S\left(\sum_{k=1}^{l} \sum_{i=1, j=R_{i(k)}}^{s(k)} f\left(w_{ij(k)}^{(1)}\right) + \sum_{k=1}^{l} \sum_{i=1, j=R_{i(k)}}^{s(k)} f\left(w_{ij(k)}^{(2)}\right) + *, \right.
$$

$$
\left. \dots, \sum_{k=1}^{l} \sum_{i=1, j=R_{i(k)}}^{s(k)} f\left(w_{ij(k)}^{(l)}\right) + * \right).
$$

The values of the coordinates y_3 are defined by calculations over the vectors $w'_1(R_1), w'_2(R_2), ..., w'_l(R_l)$. The keys $R_1, R_2, ..., R_l$ are bound and changes in any of them leads to change y_3. The brute force attack on key R has a complexity equal to q^l.

Brute force attack on the vectors $\left(t_{0(k)}, \dots, t_{s(k)}\right)$ and $\left(\tau_{0(k)}, \tau_{1(k)}, \dots, \tau_{s(k)}\right)$. The brute force attack on $\left(t_{0(k)}, \dots, t_{s(k)}\right)$ is a general for the MST cryptosystems and for the calculation in the field F_q over the group center $Z(G)$ has an optimistic complexity estimation equal to q. For the proposed algorithm all calculations are executed on whole group $|A_l(n, \theta)| = q^l$ and is a such case the complexity of the brute force attack on $\left(t_{0(k)}, \dots, t_{s(k)}\right)$ and $\left(\tau_{0(k)}, \tau_{1(k)}, \dots, \tau_{s(k)}\right)$ will be equal to q^l.

Attack on the Algorithm. The attack on the implementation algorithm of the MST cryptosystem based on the generalized Suzuki 2-group is multifaceted. Practical attacks look at the features of logarithmic signatures and random coverings known to a cryptanalyst. One solution is to use aperiodic logarithmic signatures. In the new cryptosystem with homomorphic encryption, random covers are a secret for the cryptanalyst. In this case, the known attacks based on the weakness of logarithmic signatures are impossible.

Let's estimate security and keys parameters for generalized Suzuki-2 group cryptosystem. We fix a generalized Suzuki 2-group $A_l(n, \theta) = \{S(a_1, a_2, \dots, a_l) | a_i \in F_q\}$, which is defined over the field F_q, $q = 2^n$. Then for l-parametric group we achieve $K = nl$ bit cryptography. Logarithmic signature array and random covers are known parameters that are used in encryption as follows

$$\alpha_k = \left[A_{1(k)}, \dots, A_{s(k)}\right] = \left(a_{ij}\right)_k = S\left(a_{ij(k)}^{(1)}, a_{ij(k)}^{(2)}, \dots, a_{ij(k)}^{(l)}\right),$$

$$h_{(k)} = \left[h_{1(k)}, \dots, h_{s(k)}\right] = S\left(h_{ij(k)}^{(1)}, h_{ij(k)}^{(2)}, \dots, h_{ij(k)}^{(l)}\right)$$

Also, we know random cover with homomorphic encryption

$$g_{(k)} = \left[g_{1(k)}, \dots, g_{s(k)}\right] = S\left(g_{ij(k)}^{(1)}, g_{ij(k)}^{(2)}, \dots, g_{ij(k)}^{(l)}\right)$$

for $k = \overline{1, l}$.

The number of vectors in arrays α_k, $h_{(k)}$, $g_{(k)}$ is defined by the type of logarithmic signature. $\left(r_{1(k)}, \dots, r_{s(k)}\right)$ and equals to $N = \sum_{k=1}^{l} \left(r_{1(k)} + r_{2(k)} + \dots + r_{s(k)}\right)$

Since arrays α_k, $g_{(k)}$ are random and can be constructed by random bits deterministic generator from some initial vector V, then we can define α_k, $g_{(k)}$ over the vector V. Let's fix the vector length V to be equal to nl bits.

The array size $g_{(k)}$ equals to: $N_g = l \sum_{k=1}^{l} \left(r_{1(k)} + r_{2(k)} + \dots + r_{s(k)}\right)$ n-bits words.

The secret parameters of the cryptosystem include vectors t, τ, ρ:

$$t_{0(k)}, \dots, t_{s(k)} \in A_l(n, \theta) \backslash Z, \ t_{i(k)} = S(t_{i1(k)}, \dots, t_{il(k)}),$$

$$\tau_{0(k)}, \dots, \tau_{s(k)} \in A_l(n, \theta) \backslash Z, \ \tau_{i(k)} = S(\tau_{i1(k)}, \dots, \tau_{il(k)}), \ \rho = (\rho_1, \rho_2, \dots, \rho_l), \ k = \overline{1, l}.$$

The number of vectors $t_{i(k)}$, $\tau_{i(k)}$ equals to: $N_t = N_\tau = l \sum_{k=1}^{l} s(k)$ n-bits words.

The length of the vector ρ equal to nl bits.

Obviously, that N_g, N_t, N_τ depends on type of $\left(r_{1(k)}, \dots, r_{s(k)}\right)$.

Let the secrecy of cryptographic transformations be determined by K bits.

Let's define a type of $\left(r_{1(k)}, \dots, r_{s(k)}\right) = (2, \dots, 2)$, then $s(k) = n$ over the field $F(2^n)$. We get the following values

$$N_g = nl \sum_{k=1}^{l} \left(r_{1(k)} + r_{2(k)} + \dots + r_{s(k)}\right) = 2n^2l^2 = 2K^2 \text{ bit}$$

$$N_t = N_\tau = nl \sum_{k=1}^{l} s(k) = n^2 l^2 = K^2 \text{ bit}$$

The length of vectors V, ρ equals to $N_V = N_\rho = nl = K$ bit. Let's define a type of $\left(r_{1(k)}, \ldots, r_{s(k)}\right) = \left(2^8, \ldots, 2^8\right)$, $s(k) = n/8$ over field $F(2^n)$. We achieve

$$N_g = nl \sum_{k=1}^{l} \left(r_{1(k)} + r_{2(k)} + \ldots + r_{s(k)}\right) = 2^5 n^2 l^2 = 2^5 K^2 \text{ bit}$$

$$N_t = N_\tau = nl \sum_{k=1}^{l} s(k) = n^2 l^2 / 8 = 2^{-3} K^2 \text{ bit}$$

Estimated implementation costs are presented in the table below.

Memory costs for arrays of shared and secret parameters do not depend on the field $F(2^n)$ and the number of parameters of the generalized Suzuki group. Selection of field F_q and parameters of the Suzuki group will define the speed of calculations on the group and depends on the software implementation (Table 10).

Table 10. Estimated implementation costs

$K = 256$, $\left(r_{1(k)}, \ldots, r_{s(k)}\right) = (2, \ldots, 2)$			
$F(2^n)$	N_g Kbyte	$N_t(N_\tau)$, Kbyte	$N_V(N_\rho)$, bit
$F(2^8), \ldots, F(2^{256})$	4	2	256
$K = 256$, $\left(r_{1(k)}, \ldots, r_{s(k)}\right) = \left(2^8, \ldots, 2^8\right)$			
$F(2^8), \ldots, F(2^{256})$	64	0.25	256
$K = 512$, $\left(r_{1(k)}, \ldots, r_{s(k)}\right) = \left(2^8, \ldots, 2^8\right)$			
$F(2^8), \ldots, F(2^{512})$	64	32	512
$K = 512$, $\left(r_{1(k)}, \ldots, r_{s(k)}\right) = \left(2^8, \ldots, 2^8\right)$			
$F(2^8), \ldots, F(2^{512})$	1024	8	512

4 Conclusions

Generalized Suzuki 2-groups are multiparameter groups and may have an arbitrarily large order. MST cryptosystems based on generalized Suzuki 2-group have an advantage over other schemes implementations in secrecy and realization. We can build a highly secure cryptosystem with group computation in a small finite field. Applying homomorphic encryption to random coverings in a logarithmic signature provides protection against known attacks on logarithmic signature implementations. To build a cryptosystem, you can use secure logarithmic signatures of a simple design, which leads to low costs for the general parameters of the cryptosystem. The proposed cryptosystem with homomorphic encryption is a good candidate for post-quantum cryptography.

Acknowledgements. This publication is based on work supported by a grant from the U.S. Civilian Research & Development Foundation (CRDF Global).

References

1. Wagner, N.R., Magyarik, M.R.: A public-key cryptosystem based on the word problem. In: Blakley, G.R., Chaum, D. (eds.) CRYPTO 1984. LNCS, vol. 196, pp. 19–36. Springer, Heidelberg (1985). https://doi.org/10.1007/3-540-39568-7_3
2. Magliveras, S.S.: A cryptosystem from logarithmic signatures of finite groups. In: Proceedings of the 29th Midwest Symposium on Circuits and Systems, pp. 972–975. Elsevier Publishing, Amsterdam (1986)
3. Lempken, W., Magliveras, S.S., van Trung, T., Wei, W.: A public key cryptosystem based on non-abelian finite groups. J. Cryptol. **22**, 62–74 (2009)
4. Magliveras, S.S., Svaba, P., van Trung, T., et al.: On the security of a realization of cryptosystem MST3. Tatra Mt. Math. Publ. **41**, 1–13 (2008)
5. Svaba, P., van Trung, T.: Public key cryptosystem MST3 cryptanalysis and realization. J. Math. Cryptol. **4**(3), 271–315 (2010)
6. van Trung, T.: Construction of strongly aperiodic logarithmic signatures. J. Math. Cryptol. **12**(1), 23–35 (2018)
7. Khalimov, G., Kotukh, Y., Khalimova, S.: MST3 cryptosystem based on the automorphism group of the Hermitian function field. In: IEEE International Scientific-Practical Conference: Problems of Infocommunications Science and Technology, PIC S and T 2019 - Proceedings, pp. 865–868 (2019)
8. Khalimov, G., Kotukh, Y., Khalimova, S.: MST3 cryptosystem based on a generalized Suzuki 2 – Groups. CEUR Workshop Proc. **2711**, 1–15 (2020)
9. Khalimov, G., Kotukh, Y., Khalimova, S.: Encryption scheme based on the automorphism group of the Ree function field. In: 2020 7th International Conference on Internet of Things: Systems, Management and Security, IOTSMS 2020, 9340192 (2020)
10. Hanaki, A.: A condition on lengths of conjugacy classes and character degrees Osaka J. Math. **33**, 207–216 (1996)
11. P. Svaba, "Covers and logarithmic signatures of finite groups in cryptography", Dissertation, https://bit.ly/2Ws2D24

End-to-End Security Scheme for E-Health Systems Using DNA-Based ECC

Sanaz Rahimi Moosavi[(✉)] [iD] and Arman Izadifar

California State University Dominguez Hills (CSUDH), Carson, CA 90747, USA
srahimimoosavi@csudh.edu

Abstract. Today, the amount of data produced and stored in computing Internet of Things (IoT) devices is growing. Massive volumes of sensitive information are exchanged between these devices making it critical to ensure the security of these data. Cryptography is a widely used method for ensuring data security. Many lightweight cryptographic algorithms have been developed to address the limitations of resources on the IoT devices. Such devices have limited processing capabilities in terms of memory, processing power, storage, etc. The primary goal of exploiting cryptographic technique is to send data from the sender to the receiver in the most secure way to prevent eavesdropping of the content of the original data. In this paper, we propose an end-to-end security scheme for IoT system. The proposed scheme consists of (i) a secure and efficient mutual authentication scheme based on the Elliptic Curve Cryptography (ECC) and the Quark lightweight hash design, and (ii) a secure end-to-end communication based on Deoxyribonucleic Acid (DNA) and ECC. DNA Cryptography is the cryptographic technique to encrypt and decrypt the original data using DNA sequences based on its biological processes. It is a novel technique to hide data from unauthorized access with the help of DNA. The security analysis of the proposed scheme reveals that it is secure against the relevant threat models and provides a higher security level than the existing related work in the literature.

Keywords: Deoxyribonucleic Acid (DNA) · Elliptic curve cryptography · E-health system · End-to-end security

1 Introduction

The Internet of Things (IoT) is a new paradigm for modern pervasive wireless communications that connects a wide range of physical devices via the Internet to collect and exchange data. A healthcare IoT network consists of smart sensors, wearable/implantable health devices, and medical instruments that can remotely monitor a patient's health. Among the major areas of concern in healthcare IoT are patient's security and privacy. In this regard, remote health caregiver (end-user) authentication and authorization, as well as end-to-end data protection, are critical requirements to prevent eavesdropping on sensitive medical data

© The Author(s) 2022
S.-Y. Chang et al. (Eds.): SVCC 2021, CCIS 1536, pp. 77–89, 2022.
https://doi.org/10.1007/978-3-030-96057-5_6

or malicious triggering of specific tasks [1]. As humans are directly involved in healthcare IoT applications, robust and secure data communication among healthcare sensors, actuators, patients, and caregivers is critical. Cryptography is defined as the technique that applies logic and mathematics to keep and send information in the coding style and through a secured format so that only the intended receiver can read and translate the meaning. State-of-the-art security and protection mechanisms, such as existing cryptographic solutions, secure protocols, and privacy assurance, cannot be re-used in healthcare IoT systems due to resource constraints, security level requirements, and system architecture [2]. Strong network security infrastructures for short and long-range communication are required to mitigate the aforementioned risks.

Unlike symmetric ciphers, which uses the same secret key to encrypt and decrypt sensitive data, asymmetric ciphers, also known as public-key cryptography or public-key encryption, uses mathematically linked public- and private-key pairs to encrypt and decrypt sensitive data sent to and received from senders and recipients. An important advantage of asymmetric ciphers over symmetric ciphers is that no secret channel is necessary for the exchange of the public key. The receiver needs only to be assured of the authenticity of the public key. Due to its superiority in generating a powerful encryption mechanism with small key sizes, ECC is widely used in constrained environments for asymmetric cryptography. ECC improves device performance while decreasing power consumption, making it suitable for a wide range of applications, including healthcare IoT. On the other hand, Deoxyribonucleic Acid (DNA) cryptography can be defined as a technique of hiding data in terms of DNA sequence. In the cryptographic technique, each letter of the alphabet is converted into a different combination of the four bases which make up the human's DNA [4,5]. DNA cryptography is a rapidly developing technology that is based on DNA computing concepts. Beside of the huge parallelism, DNA molecules also have massive storage capacity. A gram of DNA molecules consist of 10^{21} DNA bases which is nearly about 10^8 tera-byte [6]. As a result, it can be concluded that a few grams of DNA can contain all of the world's data [7]. These benefits of DNA computation inspire the concept of DNA cryptography. Prof. Leonard Adleman, also known as the 'A' of the RSA algorithm, is regarded as the father of DNA computing [3].

In this paper, an end-to-end security scheme for healthcare IoT systems is proposed. The main contributions of this work are twofold. First, we present our end-to-end security solution for e-health systems. In this regard, we exploit the DNA and ECC cryptography techniques. Second, we analyze the characteristics of the proposed scheme in terms of security. The security analysis of the scheme demonstrates that it is secure against the relevant threat models and offers a higher security level than the existing related work in the literature.

The remainder of this paper is organized as follows: Sect. 2 provides an overview of related work. The end-to-end security scheme for e-health systems using DNA-based ECC is presented in Sect. 3. Section 4 provides a comprehensive security analysis of our scheme. Finally, Sect. 5 concludes the paper.

2 Related Work

Roy *et al.* [8] devised a method based on DNA synthesis to improve key generation. The encryption and decryption processes are optimized by this system. A first level key and an encryption algorithm are used to convert plain text to primary cipher text. The concept of a second level key is introduced, enhancing the security of this technique. The second level private key strengthens the cipher text by adding primers and intron positions. Excellent results are obtained after analyzing the proposed method against brute force attacks. The hacker would need more than a half-year to decrypt the cipher text using a modern computer. This method has a high level of time and space complexity. Shinde *et al.* [9] proposed a new DNA-based cryptography technique. The method combines traditional cryptographic techniques with novel approaches to improve data security. The plaintext is first converted into an ASCII value, and then into binary strings. The binary strings are then converted to hexadecimal values, and a 128 bit key is generated using the MD5 algorithm. This key is converted into a 32-character hexadecimal string that is mapped to 16 dynamic values. The binary values are encoded using a mapping table. Following encoding, some mathematical and logical operations are carried out. This technique is both quick and efficient. However, he security offered in this algorithm is not suitable for healthcare IoT systems. Gogte *et al.* [10] presented a new type of DNA cryptography system for secure communication based on quantum cryptography. Quantum cryptography is a new security technique in which two parties communicate using a quantum channel. Its foundations are Heisenberg's uncertainty principle and the no-cloning theorem. Initially, a simulation of quantum key exchange and authentication is carried out. This is followed by the use of a DNA-based algorithm. The DNA encryption algorithm employs a symmetric block cipher with a 128 bit key as input. The method is secure against man-in-the-middle attacks, eavesdropping, replay attacks, packet sniffing, and spoofing. However, the technique is heavy-weight to be implemented for the resource-constrained e-health systems. A DNA cryptographic algorithm was proposed by Zhang *et al.* [11]. The method is based on the assembly of DNA fragments. The authors' algorithm incorporates DNA digital coding, DNA molecular keys, and some software techniques. This method is based on the concept of symmetric key cryptography. The encryption mechanism in this case is accomplished through the use of DNA digital coding. The main challenge of this algorithm is the implementation of the DNA molecular key. Ibrahim *et al.* [12] proposed using double DNA sequences to improve the security of data hiding. The main idea behind the scheme's design is to encrypt secret messages to ensure security and robustness. The encrypted message is tucked away in a different DNA reference sequence. Overall, a new data concealment algorithm based on DNA sequences has been suggested. The hiding of data in repeated characters in this scheme reduces the rate of modification. However, in this approach, if the attacker manages to obtain the secret message then the method is broken.

3   End-to-End Security Through DNA-Based ECC

In this section, we present our end-to-end security scheme for healthcare IoT systems. The proposed scheme consists of (i) ECC-based mutual authentication and authorization, (ii) DNA-ECC-based encryption. Our scheme offers the first-level of security through ECC algorithm requiring smaller key size and less computation overhead. The second level of security is provided by the use of a low computation DNA-ECC cryptosystem. The structure of a DNA-based cryptography techniques is shown in Fig. 1.

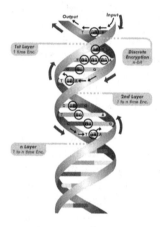

Fig. 1. DNA-based cryptography technique

3.1   ECC-Based Mutual Authentication and Authorization

This section describes our ECC-based mutual authentication and authorization scheme, which meets the security requirements of a healthcare IoT system. A mutual authentication scheme allows the communicating parties, the medical sensor device and the health care provider, to verify and ensure each other's identities. The scheme is divided into two phases: (i) health caregiver authentication and (ii) medical sensor device identification and verification. Elliptic curve cryptography (ECC) was firstly proposed by Victor Miller and Neal Koblitz [13]. It is a type of public key encryption system used to generate smaller, faster, and more efficient cryptographic keys. In contrast to the RSA algorithm, which is based on large prime numbers, keys in the ECC are generated using the elliptic curve equation's parameters. The encryption functionality provided by ECC requires fewer resources than RSA or other public key algorithms. In general, the longer the key, the better the protection for any system. However, when compared to RSA, ECC can provide comparable protection with a smaller key size. As a result, the ECC's resources must perform fewer mathematical computations. The security level of ECC can be achieved with a 164-bit key, whereas other systems require a 1024-bit key. Furthermore, the security of ECC is based on the difficulty of the Elliptic Curve Discrete Logarithm Problem (ECDLP), and because the computation of DLP problems is not easy, it prevents an adversary from easily breaking the ECDLP. An ECC E [14] over a finite field $\mathbb{F}_p$ includes all points $(x, y) \in \mathbb{F}_p \times \mathbb{F}_p$ which fulfill an equation of the form $Y^2 + a_1 XY + a_3 Y = X^3 + a_2 X^2\ a_4 X + a_6$ with $a_i \in \mathbb{F}_p$, whose discriminant is non-zero, accompanied by the point at infinity. Then, an Elliptic Curve $E (\mathbb{F}_p)$ over $\mathbb{F}_p$ defined by parameters $a, b \in \mathbb{F}_p$ made up of serious of points $P = (x, y)$ for $x, y \in \mathbb{F}_p$ to the equation:

$$y^2 \equiv x^3 + ax + b \quad (mod\ p) \tag{1}$$

The mentioned equation $y^2 \equiv x^3 + ax + b \quad (mod\ p)$ is called the description of the equation $E (\mathbb{F}_p)$ for a certain point $p = (xp, yp)$. Here, xp is entitled as the x-coordinate of P, and yp is called the y-coordinate of P. The number of point on $E (\mathbb{F}_p)$ represents as $\# E (\mathbb{F}_p)$ and:

$$p + 1 - 2\sqrt{p} \leq \# E (\mathbb{F}_p) \leq p + 1 + 2\sqrt{p} \tag{2}$$

The elliptic curve version of the Digital Signature Algorithm is known as the Elliptic Curve Digital Signature Algorithm (ECDSA) (DSA). The ECDSA is a modified version of the DSA and RSA that works with Elliptic Curve groups. The proposed ECDSA not only offers smaller key sizes for the same security level, but it also significantly improves ECC generation and authentication techniques. The diagram of elliptic curve is shown in Fig. 2. The proposed ECC-based mutual authentication scheme establishes a secure channel between the sensor and the caregiver, allowing them to communicate securely and efficiently. Before delving into the phases, we go over the parameters and notations used in the scheme.

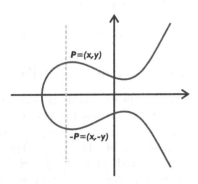

Fig. 2. Elliptic curve diagram

- G: a group of order q on an elliptic curve having the order n,
- P: a primitive element or the base point of G,
- sp_1, sp_2: each sensor keeps two secret points $sp_1, sp_2 \in E(F_g)$, which will change over time. These secret points will be varied each time the sensor is successfully identified,

- ID_n: the sensor's identification number or ID,
- sp_3: each end-user keeps a secret point $sp_3 \in Z_n$, which will change over time. This secret point will be varied each time the end-user is successfully authenticated,
- $ID_k = sp_3.P$: the end-user's public key,
- rn, j_1, j_2: random numbers in Z_n,
- h: a lightweight hash function,
- (x, y): a signature generated by the sensor in its identification phase.

3.2   Health Caregiver Authentication Phase

The Health Caregiver authentication phase of our scheme is based on the Elliptic Curve Discrete Logarithm Problem (ECDLP) [13]. In this phase, the caregiver is assigned with a random number $rs_1 \in Z_n$ and its public key is computed as $R_1 = rs_1.P$. Next, the caregiver initializes its counter value j_1 to one and sends both R_1 and j_1 to the medical sensor. It then increments the value j_1 by rn_1. Upon receiving the message, the sensor checks whether j_2 (which is initialized to zero) is greater than j_1. If the condition holds, it replaces j_2 by j_1 and selects a random number $rn_2 \in Z_n$. Then, the sensor computes:

$$rn_3 = X(rn_2.P) * Y(R_1) \tag{3}$$

where $*$ is a non-algebraic operation over the abscissa of $(rn_2.P)$ and the ordinate of R_1 and it sends the value rn_3 to the caregiver. After receiving rn_3, the caregiver computes R_2 and sends this value to the sensor. Finally, if the following equation holds, the sensor verifies that the caregiver is authentic.

$$R_2 = rn_1.ID_n + rn_3.sp_3 \tag{4}$$

$$(R_2 - rn_1.ID_n)rn_3^{-1}.P = ID_k \tag{5}$$

3.3   Medical Sensor Authentication and Verification Phase

Our scheme's medical sensor identification and verification phase is based on the Elliptic Curve Digital Signature Algorithm (ECDSA) using Quark lightweight hash design. Quark is one of the most efficient lightweight hash designs and it was first proposed by Aumasson *et al.* [14]. Quark lightweight hash is based on non-linear Boolean functions and bit shift registers. As a result, not only is its implementation feasible, but the circuit area requirements of this hash design are ideal for implantable medical devices. A digital signature also provides identification, integrity, and non-repudiation. Because of resource constraints and the delicate use cases of healthcare IoT systems, lightweight cryptographic hash designs must be carefully considered. As a result, we use the D-Quark in our proposed ECC-based medical sensor authentication (i.e., one of the flavors of Quark) lightweight hash design rather than the general purpose hash designs. In the sensor identification phase of our scheme, the sensor's initial secret point

Table 1. DNA nucleotide to binary and decimal conversion

DNA nucleotide base	Binary equivalent	Decimal equivalent
Adenine (A)	00	10
Thymine(T)	01	20
Guanine(G)	10	30
Cytosine(C)	11	40

is $sp_1 \in E(F_g)$ from which the next secret point sp_2 and ID_n will be computed. To generate the second secret point, the sensor computes:

$$s_2 = f(X(s_1)).P \tag{6}$$

Obtaining the first secret point from the second is difficult, as it requires the computation of an ECDLP. Since the second key is generated from the second key, our scheme provides forward security. For the sake of efficiency, the function f should be selected in a manner that avoids large hamming weights for sp_2, assuring that the computation of $sp_2.P$ will be fast without compromising security [12]. Once the generation of the second secret point sp_2 is done, the sensor selects a random integer $k \in Z_g$ and computes a curve point $(d, c) = k.G$. To send its digital signed message (x, y) to the back-end system, the sensor computes $d = x \mod n$. If $x = 0$, the sensor starts to select a another random number $k \in Z_g$ and computes the next curve point and its ID as:

$$ID_n = Mb(X(sp_1)) * Mb(X(sp_2)).P \tag{7}$$

where Mb will output some middle bits of the input values. The operand * is a non-algebraic operation $\in F_g$ done over the abscissa of the first and the second secret points. Then, the sensor computes the following equation:

$$l = k(hash(ID_n) + X(sp_1).x) \tag{8}$$

If the computed $y = 0$, the sensor will start the algorithm by selecting another random integer e. Finally, the sensor sends the computed values (x, y) and (ID_n) to the back-end system. Algorithm 2 shows the pseudocode of the sensor identification phase of the proposed scheme. To verify the sensor is authentic the beck-end system selects a random integer $rn_s \in Z_n$ and it computes its public key $p_r = rn_s.P$ for $j \in [1, n-1]$, the back-end system checks whether $x, y \in Z_n$. If the result is valid, the back-end system calculates $h = Hash(ID_n)$, where Hash is the same Quark lightweight hash function that is used in the previous phase to generate the sensor's signature. Once the hash value of (ID_n) is computed, the back-end selects the leftmost bit of h and denotes it as z. Then, the back-end calculates the values U, m_1, m_2. Based on the calculated values, the back-end system computes the curve point as:

$$(x, y) = m_1.P + p_r \tag{9}$$

Finally, the back-end system will accept the sensor's signature as a valid one if the equation $r = x \bmod n$ holds.

3.4   DNA-Based ECC Cryptography

DNA cryptography is a method of concealing data in terms of DNA sequence. Each letter of the alphabet is converted into a different combination of the four bases that make up human DNA in the cryptographic technique. DNA cryptography is a rapidly developing technology that is based on DNA computing concepts. Inside the tiny nuclei of living cells, DNA stores a massive amount of information. It contains all of the instructions required to create every living creature on the planet. The main advantages of DNA computation are miniaturization and parallelism, which are not available in conventional silicon-based machines. With its unique data structure and ability to perform many parallel operations, DNA allows one to view a computational problem from a new perspective. The following are the benefits of using DNA cryptography:

Algorithm 1. DNA-ECC Cryptographic Algorithm

Input: Plaintext (P), number of bits of DNA sequence segment (k), known DNA sequence (D).
Output: Ciphertext (C)
Global Variables : ECC points, which are denoted as $(x_1, y_1), (x_2, y_2), ..., (x_n, y_n)$, an auxiliary base parameter k for which both entities need to agree upon this; **Global Constants:** $Tasks$: Vector of ECC **Body:**

1: Input $Plaintext$
2: Convert $Plaintext$ into Binary P'
3: Convert D into Binary D'
4: Segment D' with k bit in a segment
5: Insert each bits of P' into the beginning of each segment of D'
6: Concatenate segments of D'
7: Convert each character of the DNA Nucleotide into Numbers as $10(00), T = 20(01), G = 30(10), C = 40(11);$
8: Koblitz Method: Pick an elliptic curve $ECC = (a, b).$
9: Each number mk, takes $x = mk + 1$ and tries to solve for $y.$
10: **while** $y \neq solved$ **do**
11:     **for each** $x \in Tasks$ **do**
12:         $x \leftarrow mk + k - 1;$
13:         Take the point (x, y) and covert m into a point on the $ECC;$
14:     **end for**
15: **end while**

---

1. *Power Requirements:* While the computation is taking place, no power is required for DNA computing. Chemical bonds, which are the building blocks of DNA, form without the assistance of an outside source of energy. The power requirements of traditional computers are incomparable.

2. *Speed:* Conventional computers have a peak performance of about 100 MIPS (millions of instruction per second). Combining DNA strands as demonstrated by Adleman made computations equivalent to 10^9 or better, arguably over 100 times faster than the fastest computer.
3. *Storage Requirements:* Memory is stored in DNA at a density of about 1 bit per cubic nanometer, where conventional storage media requires 10^{12} cubic nanometers to store 1 bit.

A simple mechanism of transmitting two related messages while concealing the message is insufficient to prevent an attacker from breaking the code. DNA cryptography has a unique advantage for secure data storage, authentication, digital signatures, steganography, and other applications. DNA strands are long polymers made up of millions of linked nucleotides. As Algorithm 1 indicates, these nucleotides are made up of four nitrogen bases, a five-carbon sugar, and a phosphate group. The nucleotides that make up these polymers are named after the nitrogen base in which they are composed: Adenine (A), Cytosine (C), Guanine (G) and Thymine (T). This means we can utilize this 4 letter alphabet $\Sigma = \{A, G, C, T\}$ to encode information, which is more than enough considering that an electronic computer needs only two digits, 1 and 0, for the same purpose. Three DNA cryptography methods are used in this cryptosystem. They are (i) the insertion method, (ii) the substitution method, and (iii) the complementary pair approach. A common method of encoding and decoding is used in all of these approaches. Binary numbers are generated from the plaintext. The binary numbers are then converted to a DNA nucleotide sequence.

Algorithm 2. DNA-ECC Cryptographic Example

Input: *Plaintext Message (P):* "m"
ASCII Message: 109
Binary Message(P'): 01101101
DNA Sequence (D): TCGCAATTCGCGCTGAGTCACAATTCGCGCTGAGTCACA
ATTCGCGCTGAGTCACAATTGTGACTCAGCCGCGAATTCCTGCAGCCCCGA
ATTCCGCATTGCAGAGATAATTGTATTTAAGTGCCTAGCT.
Output: *Ciphertext (C)* **Body:** *converting plaintext to ciphertext*

- Binary DNA Sequence (D'): 01 11 10 11 10 11 01 01 00 00 10 11 01 00
- Segmented Binary DNA sequence (where k = 3): 110 100 111 010 000 101 111 110 110 1
- Insert each bits of P' into beginning of each segments of D':1-010—1-111—0-000—1-111—1-100—0-000—0-101—110—1
- Concatenate the segments of D':00001111110000001010010101011101
- Convert D' 10-00- 11-11-11- 00- 00- 00-10-10- 01-11-01- 01-11-01 to DNA nucleotide A- C- A- A- G- G- C- C- C- A- T- T- C- A- G- T
- Convert DNA nucleotide to ASCII A C A A G G C C C A T T C A G T 10 40 10 10 30 30 40 40 40 10 20 20 40 10 30 20
- Convert the ASCII of DNA nucleotide to ECC point

The following facts underpin the encoding and decoding operations. As shown in Table 1, there are four basic units in DNA that are encoded into binary in the following manner. Binary equivalent of a DNA nucleotide base Adenine (A) is 00, Thymine (T) is 01, Guanine (G) is 10, and Cytosine (C) is 11. The DNA sequences in this work are taken from a publicly available database and converted into a binary sequence. The binary DNA sequences are divided into segments, each of which contains a random number of bits greater than two. Then, each bit of binary plain text is inserted at the start of a segmented binary DNA sequence. The inserted sequences are concatenated to obtain an encoded binary sequence. The segments are needed to be concatenated again and converted to Nucleotide letter. For encryption, we use the Koblitz method to convert decimal numbers into elliptic curve points. The plaintext is represented by ECC curve points. The ECC encryption algorithm is used to encrypt these points (Algorithm 2) [16].

$$\{kG, \ Pm \ + \ k \ PB\} \tag{10}$$

where, G is the generated points, Pm is the plaintext points, k is a random number being selected by the user, and PB is the public key of the user. The ECC decryption algorithm is used to decipher the ciphertext points. The Koblitz method is used to convert deciphered points into numbers. These numbers are decoded using DNA nucleotides, and the required plaintext is obtained.

$$Pm + kPB - nB(kG) = Pm + k(nB)G - nB(kG) \tag{11}$$

4   Security Analysis of the Proposed Scheme

In this section, we analyze the security of the proposed scheme in order to verify whether the essential requirements have been satisfied.

Mutual Authentication: In the end-user authentication phase of our scheme, to verify that the end-user is legitimate, the medical sensor computes whether $(R_2 - rn_1.ID_n)rn_3^{-1}.P = ID_k$ holds or not. Similarly, to verify whether the medical sensor is authentic (based on its transmitted (ID_n) and the digital signed message), the end-user needs to checks if $r = x$ mod n holds. This is how our proposed scheme achieves mutual authentication.

Availability: In our scheme, since the sensor and the end-user change their secret points sp_1, sp_2, and sp_3 once they are successfully authenticated, it is not possible that an adversary performs a denial of service attack.

Forward Security: Here, if an adversary attempts to decrypt some of the information he has intercepted, for example the sensor's second secret key s_2, he/she cannot benefit from the gained information. Obtaining the first secret key from the second will necessitate a solution to the ECDSA, which will be difficult.

Impersonation Attack: Concerning this type of attack, we consider two different scenarios: (i) *Impersonation of the end-user*: Here, if an adversary tries to impersonate the end-user, he/she will fail. This is because if the attacker tries to impersonate as a fake health caregiver to the medical sensor, he/she must compute R_1 and at the same time try to guess the value rn_2 (which is not easily feasible). Nevertheless, without the end-user's computed value $R_2 = rn_1.ID_n + rn_3.sp_1$, the adversary cannot compute $(R_2 - rn_1.ID_n)rn_3^{-1}.P = ID_k$ to verify whether the end-user is authentic. (ii) *Impersonation of the medical sensor*: In order to impersonate the sensor in our proposed scheme, an adversary needs to have an access to the sensor's secrets sp_1 and sp_2 and as it was presented earlier in this section, the values of the secret keys cannot be acquired from the public information of the system ID_n.

Brute-Force Attack: The DNA sequences in the proposed scheme are chosen randomly from a pool of available DNA sequences. Hence, it is impossible to predict the DNA sequence used in this study. In other words, no predictive model can be used by an attacker to determine the used DNA sequence. Without knowledge of the DNA sequence, the attacker will be unable to capture the network. When each sensor is assigned multiple DNA sequences, the DNA sequence pool is formed by randomly selecting DNA sequences from a pool of thousands. Each DNA sequence in the pool is distinct from the other DNA sequences in the pool. There are currently no methods for predicting which DNA sequences are present in the pool. Using any predictive model, an attacker cannot determine the entire DNA sequence pool. As a result, without knowledge of the DNA sequence, an attacker cannot easily capture the network.

Eavesdropping: In our scheme, (i) in the sensor identification phase, if an adversary tries to guess the sensor's secrets sp_1 and sp_2, the only public information concerning it is *ID*. As it was discussed earlier, the bits of the sensor's *ID* result from a non-algebraic operation done over some middle bits of the abscissa of two different secret points sp_1 and sp_2. Thus, it is computationally unfeasible to obtain the secret from its *ID*. (ii) In the digital signature generation section, if an adversary could guess the value x, it cannot obtain the value y effortlessly. This value is also generated from a non-algebraic operation done over the abscissa of the secret point sp_1 and the value x. The gained result will be added to the hash value of ID_n and multiplied by a random number k. Such an operation cannot be easily computed by an adversary as it requires to compute the discrete logarithm problem that is not computationally feasible. For the same reason, in the end-user authentication phase, even if an adversary could guess one of the values R_1 or R_2 or rn_3, he/she still cannot easily obtain other secure information related to the end-user. Based on the discussion above, the adversary also cannot implement any *Replay Attack*.

Unauthorized Tracking of the Sensor: Here, the only public information concerning the sensor is its *ID*. In the sensor identification phase, it was shown that the value of the sensor's *ID* results from the product of a non-algebraic operation done over some middle bits of the abscissa of the first and second secret keys of the sensor. Hence, it is impossible to compute and obtain the sensor's secret keys from its current *ID*. The main reason for this is that obtaining the secret points necessitates solving the elliptic curve discrete logarithm problem. Solving the discrete logarithm problem is as difficult as solving the integer factorization problem, this problem cannot be solved easily. Thus far, there has not been any polynomial time algorithm proposed to solve discrete logarithm problems.

5   Conclusion and Future Work

In this paper, we presented a novel end-to-end security scheme for healthcare IoT systems using ECC and DNA cryptography techniques. To the best of our knowledge, previously proposed end-to-end security schemes, concerning e-health systems in general, cannot fully fulfill the essential security requirements of health-care IoT systems. The majority of the previously proposed solutions were not secure against most common attacks on healthcare IoT systems. The proposed scheme was specified and designed by employing (i) ECC and the Quark lightweight hash design to mutually authenticate and authorize medical sensors and end-users (i.e. health caregivers), and (ii) the DNA-based ECC cryptographic technique to encrypt and decrypt the health data using DNA sequences of the patients. We demonstrated that our proposed scheme is secure against the relevant attacks and provides a higher level of security than related work found in the literature. Based on the security analyses presented in this paper, we conclude that the proposed scheme has the appropriate features for use in e-health systems. We believe that our scheme is not just limited to health-care IoT systems and can also be applied to any application of IoT that requires secure and efficient end-to-end communication. Our future work will focus on performance analysis of the proposed scheme in terms of terms of communication overhead, latency, and memory footprint.

References

1. Hummen, R., Shafagh, H., Raza, S., Voig, T., Wehrle, K.: Delegation-based authentication and authorization for IP-based Internet of Things. In: IEEE International Conference on Sensing, Communication, and Networking, pp. 284–292 (2014)
2. Hung, X., Khalid, M., Sankar, R., Lee, S.: An efficient mutual authentication and access control scheme for WSN in healthcare. J. Netw. **6**(3), 355–364 (2011)
3. Adleman, L.M.: Molecular computation of solutions to combinatorial problems. Science **266**(5187), 1021–1025 (1994)
4. Akiwate, B., Parthiban, L.: A dynamic DNA for key-based cryptography. In: IEEE International Conference on Computational Techniques, Electronics and Mechanical Systems, pp. 223–227 (2018)
5. Rafiul, M., Rokibul, K., Akber, A., Morimoto, Y.: A DNA cryptographic technique based on dynamic DNA encoding and asymmetric cryptosystem. In: International Conference on Networking, Systems and Security, pp. 1–8 (2017)

6. Pradeeksha, A., Sathyapriya, S.: Design and implementation of DNA based cryptographic algorithm. In: IEEE International Conference on Devices, Circuits and Systems, pp. 299–302 (2020)
7. Zebari, D., Haron, H., Zeebaree, S., Zeebaree, D.: Multi-level of DNA encryption technique based on DNA arithmetic and biological operations. In: IEEE International Conference on Advanced Science and Engineering, pp. 312–317 (2018)
8. Chakraborty, R., Rakshit, G., Roy, B.: Enhanced key generation scheme based on cryptography with DNA logic. Int. J. Inf. Commun. Technol. Res. **1**(8), 370–374 (2011)
9. Gehlot, L., Shinde, R.: A survey on DNA-based cryptography. Int. J. Adv. Res. Comput. Eng. Technol. **5**(1), 107–110 (2016)
10. Gogte, S., Nemade, T., Nalawade, P., Pawar, S.: Simulation of quantum cryptography and use of DNA based algorithm for secure communication. J. Comput. Eng. **11**(2), 64–71 (2013)
11. Fu, B., Zhang, Y., Zhang, X.: DNA cryptography based on DNA fragment assembly. IEEE Int. Conf. Inf. Sci. Digital Content Technol. **1**, 179–182 (2012)
12. Abdelkader, H., Ibrahim, F., Moussa, M.: Enhancing the security of data hiding using double DNA sequences. In: Industry Academia Collaboration Conference (2015)
13. Koblitz, N.: Elliptic curve cryptosystems. Math. Comput. **A8**, 203–209 (1987)
14. Miller, V.S.: Use of elliptic curves in cryptography. In: Williams, H.C. (ed.) CRYPTO 1985. LNCS, vol. 218, pp. 417–426. Springer, Heidelberg (1986). https://doi.org/10.1007/3-540-39799-X_31
15. Aumasson, J., Henzen, L., Meier, W.: QUARK: a lightweight hash. J. Crypt. **26**(2), 313–339 (2013)
16. Vijayakumar, P., Vijayalakshmi, V., Zayaraz, G.: DNA computing-based elliptic curve cryptography. J. Comput. Appl. **36**(4), 1–4 (2011)

A Comprehensive Analysis
of Chaos-Based Secure Systems

Ava Hedayatipour[(✉)], Ravi Monani, Amin Rezaei, Mehrdad Aliasgari,
and Hossein Sayadi

College of Engineering, California State University Long Beach, Long Beach, USA
ava.hedayatipour@csulb.edu

Abstract. Chaos is a deterministic phenomenon that emerges under
certain conditions in a nonlinear dynamic system when the trajectories
of the state variables become periodic and highly sensitive to the initial conditions. Chaotic systems are flexible, and it has been shown that
communication is possible using parametric feedback control. Chaos synchronization is the basis of using chaos in communication. Chaos synchronization refers to the characteristic that the trajectories of two identical
chaotic systems, each with its own unique initial conditions, converge
over time.

In this paper, data extraction is performed on different chaotic equations implemented as circuits. Lorenz is the base system implemented
in this paper, followed by Modified Lorenz, Chua's, Lü's, and Rössler
systems. Additionally, more recent systems (e.g., SprottD Attractor) are
included in the data extraction process. The robust system implementations provide an alternative to software chaos and architectures, and will
further reduce the required power and area. These chaotic systems serve
as alternatives for quantum era computing, which will cause synchronous
and asynchronous techniques to fail. The data extracted organize different modes of chaos implementation based on the ease of their fabrication
in integrated circuits. Performance metrics including power consumption,
area, design load, noise, and robustness to process and temperature variant are extracted for each system to demonstrate a figure of merit. The
figure of merit showcases chaos equations fitting to be implemented as a
transmitter/receiver with a mode of chaotic ciphering in communication.

Keywords: Chaos · Synchronization · Lorenz · CMOS · Gm-C filter

1 Introduction

Chaos is a deterministic phenomenon that emerges under certain conditions in
a nonlinear dynamic system when the trajectories of the state variable/variables
become aperiodic and highly sensitive to the initial conditions. In 1963, Lorenz
presented the first well known chaotic system, marking the beginning of chaos
theory, a branch of non-linear system theory which has been studied intensively
in recent years [1].

© Springer Nature Switzerland AG 2022
S.-Y. Chang et al. (Eds.): SVCC 2021, CCIS 1536, pp. 90–105, 2022.
https://doi.org/10.1007/978-3-030-96057-5_7

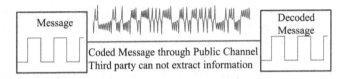

Fig. 1. Data is encrypted and transmitted over a public channel to the receiver, where it is decrypted before further processing.

Chaos can be defined as the unpredictability of a deterministic system which is highly dependent on its initial conditions. Chaos synchronization refers to the characteristic that the trajectories of two identical chaotic systems, each with its own unique initial condition, converge over time. Figure 1 shows an overview of a chaotic encryption system. The input signal is the raw unencrypted data that is scrambled by the chaotic transmitter before being transmitted over the public channel. The public channel can be wireless as in body sensor networks or wired as in a power grid.

Chaotic systems are flexible and can be utilized for communication using parametric feedback control [2]. Lorenz-based chaotic circuits can be synchronized for communication [1]. Instead of conventional frequency synthesizers, chaos generators can be used as communication carriers. Here, the digital information modulates the chaotic signal causing the digital signal to be transmitted as a chaotic spectral signal that looks like noise to a third party.

In this paper, different modes of chaotic equations suitable for communication are implemented and simulated. The robustness of system implementations are examined and their reduction of power and area compared to software implementation is explored. The data extracted organizes these different modes of chaos implementation based on the ease of fabrication in integrated circuits. For the purpose of comprehensive analysis, various performance metrics including power consumption, area, design load, frequency range, noise, and robustness to process and temperature variant are extracted and compared for each system to demonstrate a figure of merit.

2   Chaotic Ciphering of Communication

Though chaotic communication has been known for decades, with the commercialization of computers, asymmetric and symmetric key encryption Cryptography has become a fundamental part of communication between devices such as cars, implanted medical devices, and internet of things devices (IoTs). However, commonly used cryptosystems that are used in our everyday devices are expected to fail once large quantum computers exist. Quantum computing, first proposed based on a model of the Turing machine [3], originated in the 1980s based on complex phenomenons relating to quantum-mechanical physics such as superposition, the uncertainty principle, wave particle duality, and entanglement to perform computation. Later, it was suggested that computers performing quantum computation should be known as quantum computers. In October 2019,

Google in partnership with NASA, claimed they achieved quantum computing [4]; though this claim raised some dispute [5,6], it is still one of the most impressive milestones in quantum computing. Quantum computers can break symmetric and asymmetric cryptography keys quickly by exhaustively trying long bits of all secret keys. Therefore, methods of encryption other than symmetric and asymmetric security are gaining more importance and quickly becoming necessary.

Chaotic secure communication is advantageous in terms of having a strong real time performance, but real world implementation of these systems is still scarce. Research studies illustrate that the implementation of chaos theory in numerical simulations do not always perform well or expected for real world implementations. Therefore, in this paper, we focus on systems that can be implemented in hardware for various applications. Hardware based chaotic platforms can minimize area and power for a more efficient system implementation.

The Lorenz attractor, with its butterfly-like projection, is one of the first and most well-known chaotic attractors, though it has a complex equation. The main disadvantage of this system, however, lies in the necessity of using multipliers in realization of Lorenz equations, which is hard to implement. The modified Lorenz system proposed by Elwakil et al. [7] captures the essential behavior of Lorenz attractor with three differential equations and no multipliers. Notably, the Modified Lorenz system projects the "butterfly effect" and unsymmetrical Lorenz systems. As an example, to improve the stability or predictability of the Lorenz system, Stenflo and Leonov derived the following four-dimensional Lorenz-Stenflo system with four parameters. Another well-known chaos generator is Chua's chaotic system, which consists of multi-scroll chaotic oscillators derived from Chua's double-scroll equation.

Chua's equation has been implemented on a CMOS chip using $g_m - C$ modulators and non-linear resistors for the third order non linear differential equations. [8,9]. The area and power consumption were very large and the design is complex. A double scroll like chaotic oscillator was implemented using the non-linearity of CMOS inverters [10].

Current conveyor based oscillators using commercially available devices implemented Chua's equations for a master-slave communication system have been used [11]. 3−, 5−, and n-scroll attractors parameters have been approximated using real devices and integrated circuits [12,13]. Multi-scroll chaotic designs implemented using discrete components have the significant drawback of needing many external bias currents; however, V to I inverter cells which take advantage of the gate capacitance sizing can be used to address this [14].

Other than Chua's based attractors, the Lorenz equations are an alternative method of signal ciphering for communication security. The voltage equivalant of the chaotic equations, a Lorenz chaotic oscillator was fabricated back in 1999 [15], and more recently a modified Lorenz-Stnflow with reduced power consumption was.implemented as an encryption system [16]. Active control methods may be used in the system to reduce synchronization error and can be implemented using multipliers, opamps and passive components [17].

Analog circuits exhibit process, temperature, and age variations and thus present some challenges when used to implement chaotic systems. In particular, the components must have significant degrees of matching in the transmitter and receiver. It may be possible to mitigate matching issues through feedback and techniques such as using floating gates. In recent years, neural networks have been used to eliminate unwanted noise and error and train the receiver to generate the expected outputs of chaotic systems [18, 19].

Chaotic generators can also be implemented using digital systems which eliminate the matching issues found in analog circuits. However, channel noise is still a significant issues in these implementations. FPGAs have also been widely used for the implementation of chaotic systems such as Chua's system, Lü's system, Rössler's system, Chen's system, etc. [20]. However, the area and power consumption are high when using FPGA's and the implementations of these designs on integrated chips is rather rare. As the security we are targeting here mostly target portable and wearable and generally resource limited devices, use of FPGA to implement security will not be viable.

3   Design of Chaos for Integrated Circuits

Chaotic equations have been an appealing area of research for many mathematicians for more than three decades. They tried to simplify the chaotic behaviors as simple equations in order to analyze and study these behaviors. These equations aim to answer this basic question: What is the necessary and sufficient conditions for the differential equations to become chaotic?

3.1   Continuous Time

Continuous-time chaos generators are systems that can be described by nonlinear differential equations. These equations can be either differential equations (ODEs) or delay-differential equations. The positive entropy in these chaotic dynamical systems leads to continuous instability and the output being unpredictable at all times.

Chaos can be implemented using various equations. In case of all these equations, the most important block is the nonlinear element that has multiple equilibrium points, hence though the system output is unpredictable, it is bounded to "attractive regions". Integrators, sinusoidal waveform generators, delay based systems, and polynomial forms, and piecewise-linear (PWL) functions are among these non linear elements. In Table 1, different equations produce continuous chaos, along with references implementing them and their implementation based on attracting or type and function.

Table 1. Examples of continues time chaotic systems

Name	References	Equation	Scroll type	Function
Lorenz[a]	[21–23]	$x' = \lambda\sigma(y - x)$ $y' = \lambda((\beta - z)x - y)$ $z' = \lambda(xy - \rho z)$	Double scroll Multi scroll	OTA, Multiplier
Modified Lorenz[b]	[7, 24, 25]	$x' = \sigma(y - x)$ $y' = K(\beta - z) + m$ $z' = (\lvert x \rvert - \rho z)$	Double scroll Multi scroll	OTA
Lorenz-Stenflo[c]	[26, 27]	$x' = \sigma(y - x) + \lambda\omega$ $y' = (\beta - z)x - \theta y)$ $z' = xy - \epsilon z$ $\omega' = -x - \rho\omega$	Double scroll	OTA, Product
Chua[d]	[8, 9]	$x' = \sigma(y - x - f(x))$ $y' = x - y + z$ $z' = -\beta y$	Multi scroll	PWL
Rössler[e]	[28]	$x' = -y - z$ $y' = x + \sigma y$ $z' = \beta + z(x - \rho)$	Double scroll	OTA, Product
Lü[f]	[29]	$x' = \sigma(y - x)$ $y' = \beta y - xz$ $z' = -\rho z + xy$	Multi scroll	OTA, Product
SprottD	[30]	$x' = -y$ $y' = x + z$ $z' = 2y^2 + xz - a$	Multi scroll	OTA, Product

a: λ, σ, β, and ρ are parameters whose choice of value results in a chaotic system.
b: σ, β, and ρ are parameters whose choice of value results in a chaotic system.
K is a bipolar switching constant which is 1 for x $\geq$ 0 and -1 for x $<$ 0.
c: λ, σ, β, ϵ, θ and ρ are parameters whose choice of value results in a chaotic system.
d: σ, and β are parameters whose choice of value results in a chaotic system and
f(x) is a nonlinear element.
e: σ, β, and ρ are parameters whose choice of value results in a chaotic system.
f: σ, β, and ρ are parameters whose choice of value results in a chaotic system.

3.2   Discrete Time

Discrete systems can also be used to generate chaos. A discrete system is expressed as $x_{n+1} = f(C, x_n)$ that shows the next state of the system, x_{n+1} is a function of the present state, x_n, and the control parameter, C. Same as continuous time chaos, nonlinear functions are also essential here to create a chaotic map. Depending on the number of the state variables, chaotic maps are of two kinds: 1) One-dimensional maps, where deterministic equations are the only element responsible for the evolution of a single state variable, with functions such as sine map, tent map, and logistic map and 2) Multi-dimensional chaotic maps where more than one deterministic equation is needed to define the evolution of multiple state variables. In particular, Hénon map is a good example of this second category. These common functions are showcased in Table 2 as commonly used mathematical chaotic maps. Their simple mathematical expressions can be suitable for applications like FPGA-based image encryption [31]. However, it is reported that the CMOS-based compact implementation of classic chaotic maps including logistic map [32], sine map [33], and tent map [34], becomes highly hardware-hungry. As a solution to this issue, researchers have been exploring to leverage the built-in non-linearity in transistors to design simple, hardware-effective discrete maps with good chaotic properties [35–37]. Though discrete time chaos has a great potential as a base of a chaotic communication system, we will not be discussing them, as continuous time chaos is a better fit for using chaos ciphering in wearable and other resource limited systems.

Table 2. Examples of some familiar mathematical chaotic maps

Name	Mathematical expression	Parameter bounds
Logistic map [38]	$x_{n+1} = Cx_n(1 - x_n)$	$x_n = [0,1]$ $C = [0,4]$
Hénon map [39]	$\begin{cases} x_{n+1} = 1 - C_a x_n^2 + y_n \\ y_{n+1} = C_b x_n \end{cases}$	$x_n = [0,1.4]$ $C_a = [0,1.4]$ $C_b = [0,0.3]$
Sine map [40]	$x_{n+1} = C\sin(\pi x_n)$	$x_n = [0,1]$ $C = [0,1]$
Tent map [38]	$x_{n+1} = \begin{cases} Cx_n & , x_n < 0.5 \\ C(1 - x_n) & , x_n \geq 0.5 \end{cases}$	$x_n = [0,1]$ $C = [0,2]$

4 On-Chip Chaos Implementation and Simulation

Here we discuss continuous-time chaotic equations that can be described by non-linear differential equations. The steps to implement chaos as an integrated circuit are shown in Fig. 2. The first step is implementation of these equations in MATLAB, after simulations and confirmation of chaos, parameters and initial conditions that satisfy chaos are extracted. The equations are then translated to a block diagram implemented in MATLAB Simulink. The transformation from MATLAB Simulink to circuit block diagram requires implementing each block in the diagram to a low power circuit. Simulation results for each type of our equation is seen in Fig. 3. The building blocks of these systems as seen from the equations listed in Table 3 are multipliers, integrators, amplifiers, and PieceWise-Linear (PWL) functions. To extract performance parameters of each chaotic equation as an integrated chip, each block used in the chaotic system is implemented using 65 nm CMOS technology and simulated.

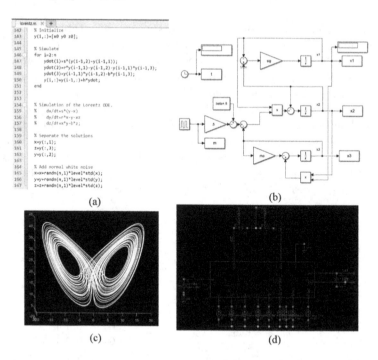

(a)   (b)

(c)   (d)

Fig. 2. The flow of implementing chaotic synchronization on chip (a) Implementing the chaotic equations in MATLAB and extracting the initial conditions and parameters needed to achieve chaos (b) Implementing the equations as MATLAB simulink block diagram (c) Simulating the chaotic equations to confirm the bounded chaos needed for ciphering of signals using chaotic synchronization (d) Implementing Simulink blocks as integrated circuits using 65 nm MOSFET technology, the block showed in the picture is an n-W powered integrator.

The first building block implemented and simulated is a low power integrator. The integrator is based on Rieger et al. [41]. This integrator, consuming power in range of nano-watts, has a very large tunable time constant without using area consuming resistors or a big capacitor. The nominal value of the time constant with a 2pF capacitor is 5 s which can also be useful for slower signals. The integrator is based on cascading of basic transconductance and transimpedance (gm - 1/gm chain). The gm blocks implemented here are 4 operational transconductance amplifiers (OTAs) with a bias current injected to them from VDD. The 1/g blocks are grounded transistors acting as voltage attenuator resistances. A chain of two gm and two 1/gm blocks, alternating, are used as an attenuator to drive the OTA and the capacitor (OTA-C). The OTA-C section consists of a current source biasing a PMOS OTA, which regulates the NMOS mirror transistors (M5-M10). This integrator is a good fit for resource limited applications that are required to consume low power and low area. Since there will be process, temperature, and voltage variations leading to an output offset on the capacitor nodes, a current source to eliminate the offset can be implemented on the last

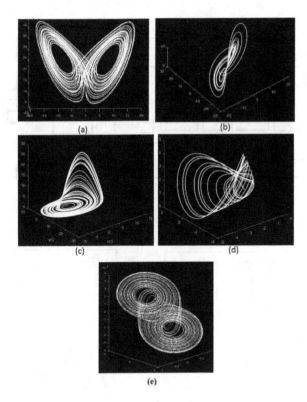

Fig. 3. Simulation results for (a) Lorenz system (b) Lü's system (c) Rössler system (d) SprottD system and (e) Chua system, The axes in the pictures are x, y and z showing the 3 states of the system.

1/gm block. The diode connected M11 is used to achieve better-balanced dc conditions. The circuit implementation of the integrator along with the simulation is seen in Fig. 4. The integrator output voltage (Vo) in this picture is simulated by extracting the response versus an square pulse as slow 4 Hz for the input.

Gilbert cell typologies are good topologies to be used as multipliers. Gilbert cells are mixers with output signals that are proportional to the product of two input signals. This cell, depicted in Fig. 5 uses eight NMOS transistors and two active loads, all are working in saturation region. There are 2 sets of differential input fed to the circuit and the top 4 transistors work as a switch that source the current in the lower part of the circuit. In the lower circuit, the signal is multiplied by the signal fed into M1–M4, and the output obtained is a differential output. To simulate this block, a sinusoidal wave and a square pulse are given as the two sets of input and as shown in the Gilbert cell simulation, the output is the multiplication of two signals. The power consumption of Gilbert cells is two orders of magnitude higher than the integrator and around 200–500 μW based on the gain implemented.

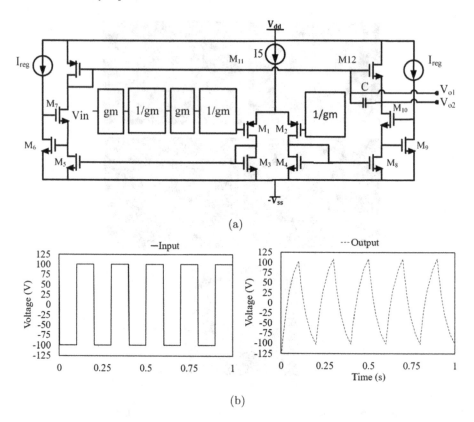

Fig. 4. (a) Circuit diagram of an integrator (b) Simulation results shown for an input pulse.

Chua's or Lü's circuit can use a CMOS implementation of the PWL as the nonlinear element of chaos. This function is constructed of various straight line segments connecting points creating custom wave-forms. PWLs are integral parts in achieving chaos since they are the limiting components when it comes to frequency. A convenient way to implement these line segments is to have a summation of simple functions. The topology shown in Fig. 6, based on Carbajal-Gomez et al. [42], can be programmed to have a break-point set by $Ioff_{in}$ and $Ioff_{out}$ and a slope that can be set by $Isat$. This implementation is also advantageous in terms of stability since it has a current mode open loop configuration. The circuit, implemented in current mode shows a better frequency response than voltage mode and is easy to implement with only few transistors.

Apart from these introduced blocks, amplifiers to determine the coefficients of the equations through gain blocks, and passive elements like resistors are common in the chaos generator circuits. It must be noted that not all the chaotic circuits discussed in literature use these blocks exclusively, but use of these blocks in this paper is to extract performance parameters using common blocks.

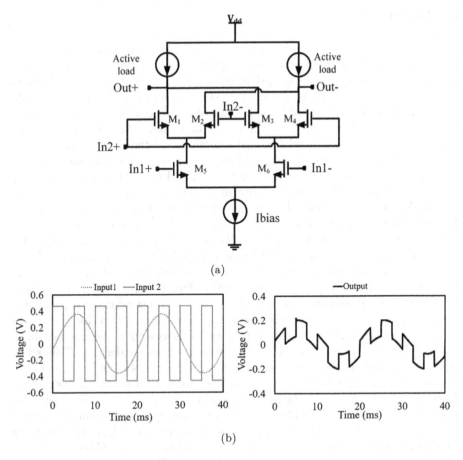

(a)

(b)

Fig. 5. (a) Circuit diagram of a multiplexer (b) Simulation results shown for multiplication of a pulse and a sinusoidal wave.

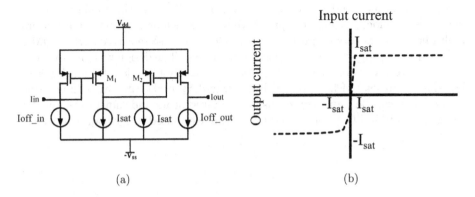

(a)   (b)

Fig. 6. (a) Circuit diagram of implementation of a piece wise-linear circuit (b) Simulation results showing the current output of a piece wise-linear circuit.

5   Discussion

To implement chaotic synchronization, different blocks introduced in the previous sections are used. In this section, performance of these equations based on their area and power consumption, sensitivity, and robustness is discussed. Chaotic synchronization is very sensitive to initial conditions, in this sense, making a small change in the initial condition of this complex, nonlinear system, produces a huge change in the behavior of the system. With slightly different initial conditions, we start with a slight difference in the results, then beyond a certain time, the system would no longer be predictable. The sensitivity of Rössler, Lü and SprottD systems can be seen in Fig. 7. Though having a big change in output with a slight change in initial conditions seems desirable as it provides better ciphering, there is indeed a trade off. Systems more sensitive toward initial conditions are also more sensitive toward process, voltage, and temperature (PVT) variations. The small changes formed in the fabrication process of the integrated chip, makes the more sensitive implementations almost impossible to synchronize as two identical systems implemented on chip will still be slightly different and posing an extremely different output. To eliminate these PVT variations, considerations for tuning circuit and post-fabrication processing are needed to contribute towards power and area consumption. Lü's system is seen to be more sensitive toward initial conditions as seen in the simulation.

Lorenz was the main equation to implement chaos for decades. This set of equations, however, needs two multipliers which are power and area consuming to be implemented on chip and can contribute to a significant DC offset. Modified Lorenz eliminates the multipliers which will reduce the power and area consumption. Chua's design also eliminates the multipliers, replacing them with a PWL circuit. The power consumption is hence proportional to the number of multipliers in the circuit as the power consummation of the Gilbert cells is 2 order of magnitude higher than that of amplifiers, PWLs and integrators.

These data are utilized to come up with a Figure of Merit (FOM), that is smaller for a better design as shown in Eq. 1. The design performance improves as we use less power and area consumption (proportional to the number of multipliers), fewer blocks (using only primitive blocks, such as amplifiers, integrators, multipliers, etc. for design purposes), and reduced noise sensitivity (the FOM is designed to be proportional with these parameters). The design will improve if we are robust to PVT variation (or if the design is easily tunable after fabrication). For the design load of Lorenz, Lü, Rössler, and SprottD, they are estimated as 1 since they can be implemented using the integrators and multipliers and the design load of Modified Lorenz and Chua's are estimated as 2 because of need to

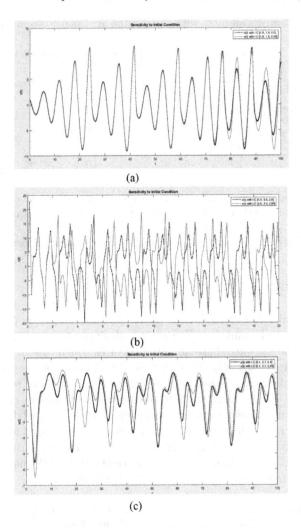

(a)

(b)

(c)

Fig. 7. Sensitivity of (a) Rössler, (b) Lü and (c) SprottD systems to initial condition.

design specialty blocks. Lorenz- Stenflo requires 4 equations to be implemented and it's design load is estimated as 2. Chua's robustness to PVT is estimated as 2 because of ease of tuning the PWL currents after the process. The detailed comparison of these systems in terms of performance is shown in Table 3.

$$FOM = \frac{Power \& AreaConsumption \times \#ofblocks \times DesignLoad \times Noise}{Robustness}$$

$$(1)$$

Table 3. Comparison table with state-of-the-art chaotic communication

Name	Based on	# of main blocks	# of multipliers	FOM
Lorenz	[23]	5	2	15
Modified Lorenz	[25]	4	0	8
Lorenz-Stenflo[a]	[26]	6	2	18
Chua	[8], [9]	4	0	4
Rössler	[28]	4	1	8
Lü[b]	[29]	5	0–2	10
SprottD[c]	[30]	5	2	7.5

a: This design has more output states leading to a more robust ciphering.
b: Various alternatives exist to implement Lü's system with one or no multiplier.
c: This design has no equilibria leading to a more robust ciphering.

6   Conclusion

In this paper Multiple Lorenz, Chua's, Lü's, and Rössler, and sprottD systems are implemented and simulated. The system implementations are considered as alternatives to the software chaos and architectures which can further reduce the power and the area overheads. Reducing power and area of these systems pave the way for the effective utilization of security at chip level for resource limited applications such as wearables, implantable devices, and Internet-of-Things (IoT) devices where security has been overlooked even in regularly used devices. Various performance metrics including power consumption, area, design load, noise, and robustness are extracted and compared for each system to demonstrate a figure of merit. The figure of merit shows the importance of reducing the use of multipliers by introducing chaotic equations that avoid using multipliers.

Acknowledgement. The simulations on this paper are done using Cadence virtuoso, supplied by Cadence university program to California State University Long Beach. This work is supported by the National Science Foundation under award No. 2131156.

References

1. Cuomo, K.M., Oppenheim, A.V., Strogatz, S.H.: Synchronization of Lorenz-based chaotic circuits with applications to communications. IEEE Trans. Circ. Syst. II Analog Digital Signal Process. **40**(10), 626–633 (1993)
2. Ott, E., Grebogi, C., Yorke, J.A.: Controlling chaotic dynamical systems. In Chaos: Soviet-American Perspective on Nonlinear Science, pp. 153–172. American Institute of Physics (1990)
3. Benioff, P.: The computer as a physical system: a microscopic quantum mechanical Hamiltonian model of computers as represented by turing machines. J. Stat. Phys. **22**(5), 563–591 (1980)

4. Arute, F., et al.: Quantum supremacy using a programmable superconducting processor. Nature **574**(7779), 505–510 (2019)

5. Kalai, G.: The argument against quantum computers. In: Quantum, Probability, Logic, pp. 399–422 (2020)

6. Smith, F.L., III.: Quantum technology hype and national security. Secur. Dialogue **5**(1), 0967010620904922 (2020)

7. Elwakil, A.S., Kennedy, M.P.: Construction of classes of circuit-independent chaotic oscillators using passive-only nonlinear devices. IEEE Trans. Circ. Syst. I Fundam. Theory Appl. **48**(3), 289–307 (2001)

8. Delgado-Restituto, M., Rodriguez-Vazquez, A.: A CMOS analog chaotic oscillator for signal encryption. In: ESSCIRC 1993: Nineteenth European Solid-State Circuits Conference, vol. 1, pp. 110–113. IEEE (1993)

9. Delgado-Restituto, M., Rodriguez-Vazquez, A., Linan, M.: A modulator/demodulator CMOS IC for chaotic encryption of audio. In: ESSCIRC1995: Twenty-First European Solid-State Circuits Conference, pp. 170–173. IEEE (1995)

10. Elwakil, A.S., Salama, K.N., Kennedy, M.P.: An equation for generating chaos and its monolithic implementation. Int. J. Bifurcat. Chaos **12**(12), 2885–2895 (2002)

11. Trejo-Guerra, R., Tlelo-Cuautle, E., Cruz-Hernández, C., Sánchez-López, C.: Chaotic communication system using Chua's oscillators realized with CCII+S. Int. J. Bifurcat. Chaos **19**(12), 4217–4226 (2009)

12. Sánchez-López, C., Trejo-Guerra, R., Munoz-Pacheco, J., Tlelo-Cuautle, E.: N-scroll chaotic attractors from saturated function series employing CCII+S. Nonlinear Dyn. **61**(1–2), 331–341 (2010)

13. Trejo-Guerra, R., et al.: Integrated circuit generating 3-and 5-scroll attractors. Commun. Nonlinear Sci. Numer. Simul. **17**(11), 4328–4335 (2012)

14. Trejo-Guerra, R., Tlelo-Cuautle, E., Jiménez-Fuentes, M., Muñoz-Pacheco, J.M., Sánchez-López, C.: Multiscroll floating gate-based integrated chaotic oscillator. Int. J. Circuit Theory Appl. **41**(8), 831–843 (2013)

15. Gonzalez, O.A., Han, G., De Gyvez, J.P., et al.: CMOS cryptosystem using a Lorenz chaotic oscillator. In: ISCAS 1999, Proceedings of the 1999 IEEE International Symposium on Circuits and Systems VLSI (Cat. No. 99CH36349), vol. 5, pp. 442–445. IEEE (1999)

16. Wu, Y.-L., Yang, C.-H., Li, Y.-S., Wu, C.-H.: Nonlinear dynamic analysis and chip implementation of a new chaotic oscillator. In: 2015 IEEE 12th International Conference on Networking, Sensing and Control, pp. 554–559. IEEE (2015)

17. Xiong, L., Lu, Y.-J., Zhang, Y.-F., Zhang, X.-G., Gupta, P.: Design and hardware implementation of a new chaotic secure communication technique. PLoS One **11**(8), e0158348 (2016)

18. Liang, C., Zhang, Q., Ma, J., Li, K.: Research on neural network chaotic encryption algorithm in wireless network security communication. EURASIP J. Wirel. Commun. Netw. **2019**(1), 1–10 (2019). https://doi.org/10.1186/s13638-019-1476-3

19. Zhang, L.: Artificial neural network model design and topology analysis for FPGA implementation of Lorenz chaotic generator. In: IEEE 30th Canadian Conference on Electrical and Computer Engineering (CCECE), pp. 1–4. IEEE (2017)

20. Tuna, M., Alçın, M., Koyuncu, İ, Fidan, C.B., Pehlivan, İ: High speed FPGA-based chaotic oscillator design. Microprocess. Microsyst. **66**, 72–80 (2019)

21. Gonzales, O.A., Han, G., De Gyvez, J.P., Sánchez-Sinencio, E.: Lorenz-based chaotic cryptosystem: a monolithic implementation. IEEE Trans. Circ. Syst. I Fundam. Theory Appl. **47**(8), 1243–1247 (2000)

22. Yu, S., Lü, J., Tang, W.K., Chen, G.: A general multiscroll Lorenz system family and its realization via digital signal processors. Chaos Interdisc. J. Nonlinear Sci. **16**(3), 033126 (2006)
23. Brown, D., Hedayatipour, A., Majumder, M.B., Rose, G.S., McFarlane, N., Materassi, D.: Practical realisation of a return map immune Lorenz-based chaotic stream cipher in circuitry. IET Comput. Digital Tech. **12**(6), 297–305 (2018)
24. Özoğuz, S., Elwakil, A.S., Kennedy, M.P.: Experimental verification of the butterfly attractor in a modified Lorenz system. Int. J. Bifurcat. Chaos **12**(07), 1627–1632 (2002)
25. Radwan, A., Soliman, A., El-Sedeek, A.: MOS realization of the modified Lorenz chaotic system. Chaos, Solitons Fractals **21**(3), 553–561 (2004)
26. Wu, Y.-L., Yang, C.-H., Wu, C.-H.: Design of initial value control for modified Lorenz-Stenflo system. Math. Probl. Eng. **2017**, 8424139 (2017)
27. Zhang, F., Chen, R., Chen, X.: Analysis of a generalized Lorenz-Stenflo equation. Complexity **2017**, 7520590 (2017)
28. Butusov, D.N., Karimov, T.I., Lizunova, I.A., Soldatkina, A.A., Popova, E.N.: Synchronization of analog and discrete rössler chaotic systems. In: IEEE Conference of Russian Young Researchers in Electrical and Electronic Engineering (EIConRus), pp. 265–270. IEEE (2017)
29. Liao, T.-L., Chen, H.-C., Peng, C.-Y., Hou, Y.-Y.: Chaos-based secure communications in biomedical information application. Electronics **10**(3), 359 (2021)
30. Wei, Z.: Dynamical behaviors of a chaotic system with no equilibria. Phys. Lett. A **376**(2), 102–108 (2011)
31. Baruah, B., Saikia, M.: An FPGA implementation of chaos based image encryption and its performance analysis. IJCSN Int. J. Comput. Sci. Netw. **5**(5), 712–720 (2016)
32. Lopez-Hernandez, J., Diaz-Mendez, A., Vazquez-Medina, R., Alejos-Palomares, R.: Analog current-mode implementation of a logistic-map based chaos generator. In: 52nd IEEE International Midwest Symposium on Circuits and Systems, pp. 812–814. IEEE (2009)
33. Farfan-Pelaez, A., Del-Moral-Hernández, E., Navarro, J., Van Noije, W.: A CMOS implementation of the sine-circle map. In: 48th Midwest Symposium on Circuits and Systems, pp. 1502–1505. IEEE (2005)
34. Callegari, S., Setti, G., Langlois, P.J.: A CMOS tailed tent map for the generation of uniformly distributed chaotic sequences. In: IEEE International Symposium on Circuits and Systems (ISCAS), vol. 2, pp. 781–784. IEEE (1997)
35. Dudek, P., Juncu, V.: Compact discrete-time chaos generator circuit. Electron. Lett. **39**(20), 1431–1432 (2003)
36. Juncu, V., Rafiei-Naeini, M., Dudek, P.: Integrated circuit implementation of a compact discrete-time chaos generator. Analog Integr. Circ. Sig. Process **46**(3), 275–280 (2006)
37. Kia, B., Mobley, K., Ditto, W.L.: An integrated circuit design for a dynamics-based reconfigurable logic block. IEEE Trans. Circuits Syst. II Express Briefs **64**(6), 715–719 (2017)
38. Zhou, Y., Hua, Z., Pun, C.-M., Chen, C.P.: Cascade chaotic system with applications. IEEE Trans. Cybern. **45**(9), 2001–2012 (2014)
39. Al-Shameri, W.F.H.: Dynamical properties of the hénon mapping. Int. J. Math. Anal. **6**(49), 2419–2430 (2012)
40. Hua, Z., Zhou, Y.: Dynamic parameter-control chaotic system. IEEE Trans. Cybern. **46**(12), 3330–3341 (2015)

41. Rieger, R., Demosthenous, A., Taylor, J.: A 230-nW 10-s time constant CMOS integrator for an adaptive nerve signal amplifier. IEEE J. Solid-State Circ. **39**(11), 1968–1975 (2004)
42. Carbajal-Gomez, V.H., Tlelo-Cuautle, E., Muñoz-Pacheco, J.M., de la Fraga, L.G., Sanchez-Lopez, C., Fernandez-Fernandez, F.V.: Optimization and CMOS design of chaotic oscillators robust to PVT variations. Integration **65**, 32–42 (2019)

Miscellaneous Security

Crowdfunding Non-fungible Tokens on the Blockchain

Sean Basu[1], Kimaya Basu[1], and Thomas H. Austin[2,3]([✉]) [iD]

[1] Monta Vista High School, Cupertino, CA, USA
[2] 0Chain Corporation, Cupertino, CA, USA
[3] San José State University, San Jose, CA, USA
thomas.austin@sjsu.edu

Abstract. Non-fungible tokens (NFTs) have been used as a way of rewarding content creators. Artists publish their works on the blockchain as NFTs, which they can then sell. The buyer of an NFT then holds ownership of a unique digital asset, which can be resold in much the same way that real-world art collectors might trade paintings.

However, while a deal of effort has been spent on selling works of art on the blockchain, very little attention has been paid to using the blockchain as a means of fundraising to help finance the artist's work in the first place. Additionally, while blockchains like Ethereum are ideal for smaller works of art, additional support is needed when the artwork is larger than is feasible to store on the blockchain.

In this paper, we propose a fundraising mechanism that will help artists to gain financial support for their initiatives, and where the backers can receive a share of the profits in exchange for their support. We discuss our prototype implementation using the SpartanGold framework. We then discuss how this system could be expanded to support large NFTs with the 0Chain blockchain, and describe how we could provide support for ongoing storage of these NFTs.

Keywords: Blockchain · Non-fungible tokens · Crowdfunding · Storage

1 Introduction

As the world moves to an online, digital retail model, there has been a struggle to find ways to reward artists and other content creators for their work.

Non-fungible tokens (NFTs) have been one proposed solution. Artists create their work and then sell it online, with ownership tracked on the blockchain to determine who owns the unique copy of the work of art.

However, while NFTs offer a model for artists to sell their work, they do not intrinsically offer a way for artists to raise funds to help them create their projects in the first place. Additionally, while NFTs offer a good model for storing smaller amounts of content on the blockchain, the cost of storing a larger work of art on the blockchain quickly becomes prohibitive. In this paper, we

S.-Y. Chang et al. (Eds.): SVCC 2021, CCIS 1536, pp. 109–125, 2022.
https://doi.org/10.1007/978-3-030-96057-5_8

highlight how an artist's project to create an NFT can be supported on the blockchain through crowdfunding. In our design, an artist posts a transaction to the blockchain advertising their project. Other clients may then contribute to the project, in exchange gaining a portion of the coins from the initial sale. Once the artist then launches the NFT on the blockchain, the funding campaign is tied to the NFT itself. The artist may then sell the NFT as they see fit, and the artist's backers are compensated automatically. For a successful artist, a history of successful projects can be an excellent form of marketing; backers can see the artist's history on the blockchain and thereby be encouraged to invest in the artist's next fundraising campaign.

To help further understanding of our design, we offer a prototype implementation using the SpartanGold framework [2]. Our implementation is available at https://github.com/taustin/spartan-gold-nft. We then consider the storage of larger NFTs, and discuss our proposed design for the 0Chain blockchain. As part of this discussion, we show how 0Chain's token-locking reward protocol [10] can be used to create an ongoing revenue stream to fund long-term NFT storage.

2   Background

Bitcoin's seminal whitepaper [12] introduced the world to the blockchain as a distributed and decentralized ledger for managing cryptocurrency. Bitcoin also included a primitive scripting language for writing programmable smart contracts. However, due to concerns about denial-of-service attacks, the power of these smart contracts was deliberately restricted.

Namecoin, a fork of Bitcoin, focused on allowing data to be stored directly on the Blockchain. In 2014, Kevin McCoy and Anil Dash used Namecoin to launch what is generally considered the first NFT [6].

Ethereum expanded upon Bitcoin's ideas to create a blockchain that supports a quasi-Turing complete virtual machine [22]. To avoid denial-of-service attacks, Ethereum's virtual machine (EVM) includes a notion of *gas*. Clients pay for their transactions by specifying a gas price; if they run out of gas, the effects of the transaction are rolled back, and the miners keep the ether used to pay for the transaction.

Ethereum's flexibility introduced the world to a wide variety of new applications for the blockchain. One popular use was the creation of ERC-20 tokens [20]. The ERC-20 specification allows a standard way for organizations to issue tokens as a fundraising mechanism. These tokens typically serve as a placeholder for the native coins on a new blockchain; once the new blockchain is launched, clients may exchange these tokens to receive an equivalent amount of native coins on the new blockchain.

While ERC-20 has been an influential design, its focus is on *fungible* tokens. There is no connection between a fungible token and any unique asset. Essentially, ERC-20 tokens act as an additional currency running on the Ethereum blockchain.

To our knowledge, the first example of a fungible token on the Ethereum blockchain was used in the design of *CryptoPunks* [4] in 2017. In this application, users trade unique cartoon characters on the blockchain. Later that same year, *CryptoKitties* was released on the Ethereum blockchain. At its height, CryptoKitties accounted for a quarter of the traffic on Ethereum's blockchain [5]. The popularity of these applications served to both highlight the power of the Ethereum blockchain, and to showcase its limitations in handling the amount of traffic generated by these applications.

The success of these applications lead to the development of two Ethereum Improvement Proposals (EIP): EIP/ERC-721 [7] provides a standard interface for non-fungible tokens; EIP/ERC-165 [15] gives a way to tag an implementation to indicate that it supports a given contract interface.

The ability to create unique tokens on a decentralized, publicly visible blockchain has lead to some initial efforts at using NFTs as a form of inventory management. Regner et al. [14] describe how NFTs can be used as part of an event ticketing system. Westerkamp et al. [21] use NFTs on Ethereum to track inventory in a manufacturing process, where "recipes" dictate how NFTs representing ingredients are consumed to produce new NFTs of the finished good. Stefanović et al. [18] describe the applications for smart contracts in handling land administration systems and real estate transfers, though the authors do not explicitly mention NFTs. Bastiaan et al. [1] describe how NFTs could be useful in real-estate management, including some discussion of early attempted applications of this work for Vermont and Ukraine. Patil [13] develops a NFT-based land registry system using government records for the Washington D.C. area. Salah et al. [16] propose a system for tracking soybeans using the Ethereum blockchain. Kim et al. [8] describe possible applications in the areas of food traceability and describe how these assets can be *tokenized.*

Alternately, NFTs have been seen as a new way to create a market for digital works of art. While CryptoPunk and CryptoKitties can been seen as initial works in this direction, additional challenges remain. Chevet [3] provides an overview of the challenges and benefits in using NFTs to reward artists, arguing that scarcity is the key property that NFTs add to the existing digital art world. Trautman [19] provides a detailed overview of the history of NFTs for virtual art, including extensive discussion of some of the highest-selling NFTs to date.

Muller et al. [11] show how their DeCoCo system can use *fungible* tokens as a mode of rewarding artists, where tokens translate to permission to access some content. While their use case is slightly different than ours, the usage of tokens to track ownership for artistic content bears a similarity to our own design.

3   Crowdfunding NFT Creation

In this section, we highlight how the blockchain can facilitate decentralized fundraising for artist projects, and also tie successful projects to the resulting NFTs. In our discussion, the artist and the backers are both assumed to be clients on the blockchain. There is also a smart contract, the *NFT Smart Contract* (NFTSC), which manages the fundraiser and records the contributions.

We assume that the backers will share the proceeds for the sale of the NFT. However, if the artist wishes to retain all funds, they may specify that when initializing the fundraiser; and non-monetary benefits from the artist must then be managed off-chain. Figure 1 shows a sequence diagram of the process.

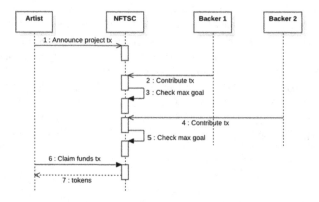

Fig. 1. Crowdfunding sequence diagram

The process for creating a new fundraiser works as follows:

1. The artist posts a transaction the the NFT Smart Contract, specifying:
   - Artist's ID.
   - Project name.
   - Project description.
   - Project ID, chosen by artist. This should be unique for the artist.
   - End date, when the fundraiser will conclude.
   - Minimum funding. If not met by the end date, the fundraiser fails.
   - Maximum funding (optional). If this amount is met or exceeded, the fundraiser ends immediately.
   - Artist share, between 0 and 1. When the NFT is eventually sold, this amount specifies what percentage of the sale goes directly to the artist.
2. Backers contribute to the project, specifying:
   - Artist's ID and the project ID.
   - Amount of tokens to contribute.
   - ID of the backer.
3. The NFT Smart Contract records the contribution. If the maximum funding goal is met, the fundraiser ends.

When the fundraiser ends, the NFT Smart Contract verifies that the funding goal has been met. If so, the contributions are recorded and the funds are transferred to the artist. Otherwise, all funds are returned to the artist. Who initiates the transaction depends on the result. If the fundraiser is successful, it is in the best interest of the artist to write the transaction in order to get access to the raised funds. Otherwise, any of the backers can call the smart contract to reclaim funds from an unsuccessful fundraiser; if there are multiple backers, this situation could result in a waiting game, where each backer hopes one of the others will bear the cost of the transaction. This issue could potentially be addressed by compensating the caller from the contributed funds.

4   Creation of the NFT on the Blockchain

After successful fundraising, the artist can use those resources to create their project. Once completed, they then call the NFT Smart Contract to release the NFT on the blockchain. Initially, the NFT is owned and controlled by the artist, though the backers' shares of the NFT are also recorded.

To create the NFT on the blockchain, the artist must write a transaction calling the NFT Smart Contract with the following information:

- artist ID.
- project ID.
- NFT data.
- NFT content hash (optional).

For smaller NFTs, the entire content might be stored on the blockchain, in which case the content hash is unnecessary. For larger NFTs, the data must be uploaded to its storage location before this transaction is written. Specifying the correct hash is the responsibility of the artist, and the blockchain miners are not expected to validate it. However, the specification of the hash allows others to verify the validity of the content off chain.

Of course, many storage schemes could be used. Section 7 shows how the NFT creation on the blockchain could be coupled with 0Chain's storage system.

Once the transaction has been written, the NFT Smart Contract records the NFT and tracks the ownership information, including details about the backers' shares.

5   Initial Sale of the NFT

With our design, the NFT Smart Contract serves as an escrow service handling the exchange of the NFT for coins on the blockchain. Once the transfer is complete, both the artist and their backers receive their coins, and the buyer receives ownership and control of the NFT. Figure 2 shows a sequence diagram for the steps in the initial sale of the NFT.

To begin this process, the artist writes a transaction calling the NFT Smart Contract. This call should specify:

- The ID of the buyer.
- The purchase price.
- The expiration of the offer.
- The NFT itself. It information should include the project ID.

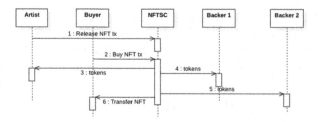

Fig. 2. NFT initial sale sequence diagram

As part of this transaction, the NFT is transferred to the ownership of the NFT Smart Contract. If the offer expires before the buyer has fulfilled the agreement terms, then the artist may call the NFT Smart Contract to reclaim its NFT.

To accept the terms of the agreement, the buyer must write to the NFT Smart Contract, specifying the NFT and transferring enough coins to meet the purchase price. If it does so before the offer expiration, the NFT is transferred to the buyer.

Once the exchange is completed, the proceeds from the sale are transferred to the artist and their backers according to the terms specified in the initial fundraising phase of the project.

6   Implementation

To help further understanding of our design, we implement our system in the SpartanGold blockchain framework. Our implementation is available at https:// github.com/taustin/spartan-gold-nft.

6.1   SpartanGold Overview

SpartanGold [2] is a JavaScript framework for simulating different blockchain designs. Its default design is roughly patterned after Bitcoin, with its miners using proof-of-work to validate transactions. However, its design is simpler, and more amenable to being easily extended with alternate designs or configurations. Transactions in SpartanGold include a `data` field, which accepts arbitrary JSON data. As a result, transactions may be extended in a variety of ways without changing the `Transaction` class. However, the logic to correctly interpret any information in the `data` field must be added to the `Block` class, described later in this section. In our design, we add a `type` field to `data`, allowing the `Block` implementation to easily add custom logic for that specific kind of transaction.

Simulations in SpartanGold can be done in a single-threaded mode, communicating through the `FakeNet` module. This approach allows for better demonstrations, with all results posted in a single window. However, the miners may instead be run in separate processes, in which case they can communicate over the network and avoid any "cheats" in the code.

A couple of differences between SpartanGold and Bitcoin should be noted. First, SpartanGold uses an account-based model. In our experience, this model is easier for students to understand than Bitcoin's UTXO model, and it simplifies many cases where we wish to tie some information to a specific account. Second, SpartanGold does not have a built-in scripting language for writing smart contracts.

Instead of using smart contracts, we must extend SpartanGold's `Block` class to handle new types of transactions. The `Block` class not only stores all transactions, but also stores any additional information that should be tracked and keeps track of the rules for validating transactions.

6.2   NFT Basic Operations

For our prototype, we first review how an NFT can be added to the SpartanGold blockchain. While NFTs are often visual works of art, in our example, we use a poet creating a new poem as an NFT. Figure 3 shows the driver for creating and transferring an NFT.

The initial code sets up a blockchain with four clients (`alice`, `storni`, `minnie`, and `mickey`), where two of the clients (`minnie` and `mickey`) are miners, and `storni` is an artist who creates an NFT. The balances for all clients are specified in the genesis block, along with the implementations for transactions and blocks. For this example, we use the standard SpartanGold `Transaction` class, but extend the `NftBlock` class with extra logic for handling NFTs.

After running for 2 s, `storni` invokes her `createNFT` method, where her NFT content is the poem "Hombre pequeñito". The code runs for an additional 3 s before `storni` then transfers the NFT to `alice`. At the 10 s mark, the blockchain terminates, and final balances are displayed. Additionally, the NFTs for `storni` and `alice` are displayed in order to show that the NFT has been successfully transferred.

The output of the program is given below. Some messages have been edited to reduce the output length, but note that `alice` has possession of the NFT at the end of program execution. The balances of the two miners have also increased, each gaining a reward of 25 gold for every block that they have produced.

```
Initial balances:
Alice:  233
Minnie: 500
Mickey: 500
Storni: 500
Mickey: found proof for block 1: 2764
Mickey: found proof for block 2: 4080

... TRIMMED FOR BREVITY ...

Mickey: found proof for block 12: 1268
***CREATING NFT***
Minnie: found proof for block 13: 9984
Mickey: found proof for block 14: 25567

... TRIMMED FOR BREVITY ...

Mickey: found proof for block 23: 54456
```

```
const {Blockchain, Miner, Transaction, FakeNet} = require('spartan-gold');
const NftClient = require('./nft-client.js');
const NftBlock = require('./nft-block.js');

let fakeNet = new FakeNet();

// Clients and miners
let alice = new NftClient({name: "Alice", net: fakeNet});
let minnie = new Miner({name: "Minnie", net: fakeNet});
let mickey = new Miner({name: "Mickey", net: fakeNet});

// Artist creating an NFT
let storni = new NftClient({name: "Alfonsina Storni", net: fakeNet});

// Creating genesis block
let genesis = Blockchain.makeGenesis({
  blockClass: NftBlock,
  transactionClass: Transaction,
  clientBalanceMap: new Map([
    [alice,233], [storni,500], [minnie,500], [mickey,500],
  ]),
});

function showBalances(client) {
  console.log(`Alice: ${client.lastBlock.balanceOf(alice.address)}`);
  console.log(`Minnie: ${client.lastBlock.balanceOf(minnie.address)}`);
  console.log(`Mickey: ${client.lastBlock.balanceOf(mickey.address)}`);
  console.log(`Storni: ${client.lastBlock.balanceOf(storni.address)}`);
}

console.log("Initial balances:");
showBalances(alice);

fakeNet.register(alice, minnie, mickey, storni);

// Miners start mining.
minnie.initialize(); mickey.initialize();

// Artist creates her NFT.
setTimeout(() => {
  console.log("***CREATING NFT***");
  storni.createNft({
    artistName: storni.name, title: "Hombre pequeñito",
    content: `
Hombre pequeñito, hombre pequeñito,
Suelta a tu canario que quiere volar...
Yo soy el canario, hombre pequeñito,
déjame saltar.`,
  });
}, 2000);

setTimeout(() => {
  let nftID = storni.getNftIds()[0];
  console.log(`***Transferring NFT ${nftID}***`);
  storni.transferNft(alice.address, nftID);
}, 5000);

// Print out the final balances after it has been running for some time.
setTimeout(() => {
  console.log();
  console.log(`Minnie has a chain of length ${minnie.currentBlock.
    chainLength}:`);
  console.log("Final balances (Alice's perspective):");
  showBalances(alice);

  console.log();
  console.log("Showing NFTs for Storni:");
  storni.showNfts(storni.address);

  console.log();
  console.log("Showing NFTs for Alice:");
  alice.showNfts(alice.address);

  process.exit(0);
}, 10000);
```

Fig. 3. SpartanGold NFT simulation

```
***Transferring NFT fc469b3105a3c89416a...
Minnie: found proof for block 24: 27051

... TRIMMED FOR BREVITY ...

Minnie: found proof for block 45: 66366

Minnie has a chain of length 46:
Final balances (Alice's perspective):
Alice:  233
Minnie: 1125
Mickey: 975
Storni: 500

Showing NFTs for Storni:

Showing NFTs for Alice:

    Alfonsina Storni's "Hombre pequeñito"
    ------------------------------------

Hombre pequeñito, hombre pequeñito,
Suelta a tu canario que quiere volar...
Yo soy el canario, hombre pequeñito,
déjame saltar.
```

```
esCrow.setContract([
   (tx) => tx.from === alice.address &&
           tx.outputs[0].amount === 150 &&
           tx.outputs[0].address === esCrow.address ,
   (tx) => tx.from === storni.address &&
           tx.data !== undefined &&
           tx.data.receiver === esCrow.address &&
           tx.data.nftID === nftID
], () => {
   esCrow.postTransaction([{ amount: 150, address: storni.
       address }]);
   esCrow.transferNft(alice.address , nftID);
});
```

Fig. 4. Setting an escrow agreement

6.3 NFT Escrow

Since SpartanGold does not have smart contracts, we must enable another way for an NFT to be transferred between clients. Our solution is to extend the SpartanGold *Client* class to create an *EscrowClient*. An *EscrowClient* can receive gold (SpartanGold's currency) or NFTs just like any other client. However, it can receive a contract of conditions that different parties agree to take through the `setContract` method.[1]

---

[1] Note that calling this method does not involve transactions on the blockchain, and it does not provide any defenses against abuse; this design simulates what would be done through a smart contract in a blockchain that supported them.

Figure 4 shows an example using the `setContract` method. The `setContract` method take an array of *conditions*, which are callback functions returning true or false. The `EscrowClient` monitors transactions, testing them against these functions. Whenever a condition is satisfied, it is removed from the list of conditions. Once the last condition is met, the `action` callback function is executed, and then the `action` itself is deleted. In the example, the contract monitors the blockchain to watch for `alice` transferring 150 gold to the escrow account and for `storni` to transfer the NFT to the escrow account. Once these actions have been completed, the escrow account posts transactions to transfer to the gold to `storni` and the NFT to `alice`.

6.4 Crowdfunding

For our implementation, the following code snippet shows how a client `storni` advertises a fundraiser, set to expire one minute after the project is posted.

```
storni.createFundraiser({
  projectName: "Un poema de amor",
  projectDescription: "Probablemente pienses que este
      canción es sobre ti, ¿no es así?",
  projectID: "1",
  endDate: Date.now() + 60000,
  minFunding: "20",
  maxFunding: "25",
  artistShare: "0.20",
});
```

The following code shows the `initFundraiser` method of the `NftClient` class. It derives a fundraiser ID from the artist's ID and the artist's choice for project ID, and then stores that fundraiser in the current block.

```
initFundraiser(artistID, projectID, {
  projectName, projectDescription, endDate, maxFunding,
      artistShare,
}) {
  let fundraiserID = this.calcFundraiserID(artistID,
      projectID);
  this.fundraisers.set(fundraiserID, {
    donations: [],
    artistID,
    projectName,
    projectDescription,
    endDate,
    maxFunding,
    artistShare,
  });
}
```

A few changes are then needed in other parts of the code. When the `createNft` method from Sect. 6.2 is invoked, the artist must specify the `projectID` matching the ID she selected during the fundraising piece. Doing so ensures that the contributors receive a share of the proceeds on the initial sale of the NFT. Of course, the artist could neglect to specify the `projectID` and keep the full sale price. However, the artist's fundraising history is on the blockchain, and a history of unfulfilled fundraisers is likely to reduce her success in fundraising again in the future. Of course, she could register additional accounts, but if she is a successful artist, changing her identity would be to her detriment.

In addition, when an NFT is sold initially, the contract with the escrow service must also be changed to reward the backers. The code below shows the modified action that could be registered for the `EscrowClient`. Note the addition of the `project` field with the relevant details of the project.

```
( project ) => {
  let payment = 450;
  let artistShare = Math.floor ( project.artistShare * payment );
  payment -= artistShare ;
  let outputs = [];
  // Giving the artist her cut.
  outputs.push( {amount: artistShare , address: storni.address });
  project.backers.forEach (({ id , amount }) => {
    let reward = Math.floor (payment * amount / project.
        totalDonations ) ;
    outputs.push( {amount: reward , address: id });
  });
  esCrow.postTransaction (outputs ) ;
  esCrow.transferNft ( alice.address , nftID ) ;
}
```

7   Storing Large NFTs on the 0Chain Blockchain

For smaller NFTs, it is feasible to store the entire NFT directly on the blockchain. However, as the storage needs increase, it becomes exceedingly expensive (or even prohibitive) to store the data directly on the blockchain. Since our focus is on providing a market for artists, we discuss how our design may be coupled with storage using the 0Chain blockchain.

In this section, we first provide a brief overview of 0Chain's design. Then we show how our system could be integrated into this blockchain.

7.1   0Chain Overview

To understand our design, a few key features of 0Chain's architecture must be understood.

0Chain advertises itself as a high performance decentralized storage network. Its *token-locking reward model* (TLRM) [10] allows for "free" transactions or other services. Instead of paying for service by transferring tokens, clients may temporarily lock their tokens (making them unavailable) in order to generate interest, acting somewhat like a bond where the interest is prepaid. That generated interest may then be given to miners or service providers. Essentially, clients pay in liquidity, but do not permanently lose their tokens.

On the 0Chain blockchain, tokens may be placed in *token pools*. A client can then give signed *markers* to other clients, allowing those other clients to draw funds from the token pools. The combination of token pools and markers is roughly analogous to banking accounts and checks.

0Chain's focus is on creating a marketplace for storage. *Blobbers* provide the storage, curated by the blockchain. Data for the blobbers is erasure coded and encrypted, ensuring that no single blobber is given an inordinate amount of power over the data that it stores. Through the use of proxy re-encryption [17], the client's data can be easily and efficiently re-encrypted for any recipient without revealing it to the blobbers themselves. Clients then pay blobbers in *read markers* and *write markers*, allowing the blobbers to draw on funds from the appropriate token pools (referred to as the read pool and write pool respectively).

7.2   Modifications Needed for Storage

The steps that we outlined in the design of our prototype implementation for the creation and sale of NFTs still apply for NFTs created for the 0Chain blockchain. However, with 0Chain, we can tie the NFTs to storage allocations directly on the blockchain. Note that the data is not stored on the blockchain itself, but the record of payment for storage and the management of the blobbers storing the data is publicly available on the blockchain.

When the artist writes a transaction to the blockchain creating the NFT, they must also specify any needed requirements for storage, such as the amount of data to be stored and the quality of service required. Additionally, they must provide a supply of ZCN (0Chain's native token) to fund the initial storage.

Since the NFT itself is not stored on the blockchain directly, a hash of the content must be stored instead. This hash allows any user accessing the NFT and its off-chain data to verify its authenticity.

When the NFT Smart Contract creates the NFT, it assigns blobbers to store the NFT based on the specifications of the artist. The tokens provided by the artist are divided between the read pool and write pool for the NFT.

Finally, an additional step is needed beyond the process listed in Sect. 4. The artist must upload the erasure coded and (optionally) encrypted data to the blobbers. As part of this interaction, the artist must send signed write markers to each blobber. These markers include a Merkle root [9] of the erasure coded data, thereby serving both as a handshake between the artist and the blobber and as a form of payment. The blobber may write a transaction to redeem these markers on the blockchain, but doing so serves as the blobber's commitment to store the data that matches the Merkle root specified by the client. A challenge protocol probabilistically ensures that the blobber is both storing the data and that it matches this agreed-upon Merkle root. The blobber is rewarded or punished depending on the results of the challenge. For more details on this challenge protocol and the format of the write markers, we refer the interested reader to Merrill et al. [10].

One significant modification is needed to 0Chain's architecture for this design. In the 0Chain ecosystem, allocations of data are permanently tied to a single account. However, with an NFT, we wish to be able to transfer control of the data corresponding to the NFT to the new owner. Adding this ability would be relatively straightforward, and could potentially lead to additional applications.

There is one significant challenge that must be addressed, however. If the data for the NFT is encrypted using proxy re-encryption [17], then the ability to generate re-encryption keys requires the original private key used to encrypt the data. This would require the blobbers to re-encrypt the data when an NFT's ownership changed, who would need to be compensated for their additional work. We will discuss that point in more detail in the next section.

When selling an NFT on 0Chain's network, the ownership of the associated data allocation must also be transferred with it. While this is not a currently supported feature, the change to do so seems fairly minor. When the allocation is transferred to the new owner, the tokens in the corresponding read and write pools remain associated with it. Therefore, the initial cost of storing the NFT will already be handled by the previous owner; handling the ongoing storage is described in the next section.

If the data for the NFT is not encrypted, no other change is needed to the process. Essentially, a storage allocation for an NFT can be handled like any other. Ideally, the allocation should be marked as read-only, thereby guaranteeing to any buyer of an NFT that the content has not changed from the original hash. While this does not guarantee that the data originally updated is correct, an auditing service could be used to review the NFT and provide a stamp of approval on the blockchain.

A more interesting case arises when the data for the NFT is encrypted. 0Chain uses proxy re-encryption. The NFT owner would first erasure code the data into separate stripes given to each storage provider, and then encrypt the stripes using its public key. When providing read access to other parties, the owner would take the receiver's public key; from that public key and the owner's own public/private key pair, the owner generates a re-encryption key. The re-encryption key is sent to the blobbers, who can re-encrypt the data as if it had been originally encrypted with the receiver's public key. The advantage of this approach is that the blobbers do not have access to the original content, but can re-encrypt the data for the receiver on the owner's behalf.

However, when transferring an NFT to a new owner, the data stored must be re-encrypted for the new owner's key pair. Fortunately, this re-encryption can be done entirely on the blobber's side. When the seller writes a transaction to transfer ownership of the NFT, they must include the valid re-encryption key.

The blobbers would need to re-encrypt their storage using the re-encryption key. However, they would need to be compensated for their work, and to ensure that the Merkle root that they have committed to storing matches the data that they are actually storing. As a result, the buyer would need to calculate the Merkle roots for each data chunk and upload matching write markers to all blobbers.

7.3   Funding Storage

For NFTs stored on the blockchain, the initial cost is high, but the NFT owner does not need to pay maintenance costs. Since data on the blockchain is permanent, it will always be available as long as the full blockchain is stored by some subset of the mining network.

In contrast, in 0Chain's ecosystem, storage is an ongoing cost. Clients pay blobbers for a period of storage; when that period ends, the client must negotiate to continue the storage contract, or else let the storage allocation expire. This model allows storage to be done more cheaply, but requires ongoing funds to maintain the NFT.

However, 0Chain's token-locking reward model can be used to create permanent storage. By locking tokens for a set period of time, the client earns additional tokens as a form of pre-paid interest. Those tokens can be spent however the client wishes, and is the basis for 0Chain's "free" transaction model.

In order to create permanent storage for an NFT, the owner needs to create an additional token pool, which we refer to as the *NFT funding pool*. The NFT funding pool may be periodically locked in order to generate an ongoing revenue stream for the NFT's write pool.

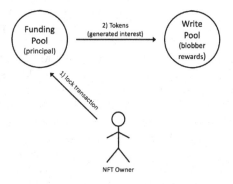

Fig. 5. Fundraising pool

For example, let's assume that the cost of storing an NFT is 20 tokens for 90 days, and that the interest rate for locking tokens is 10% for the same period. The owner can create an NFT funding pool with 200 tokens. By locking the tokens in the NFT funding pool, 20 tokens are minted and added to the write pool for the NFT. When the storage contract duration elapses, the tokens in the NFT funding pool are also unlocked, allowing the owner to relock them, thus continuing funding for the storage. Figure 5 shows a picture of this process.

Of course, this design must consider price fluctuations in the cost of storage and the value of 0Chain tokens, known as *ZCN*. Should the price of ZCN rise compared to the cost of storage, additional rewards are generated for the write pool, and could be used to offset periods where the price drops.

On the other hand, if the cost of storage drops below the amount of tokens that the NFT funding pool can generate, then the same blobbers will be unwilling to provide storage. However, other blobbers with lower quality of storage could be used as backup storage providers. We introduce the notion of *archival blobbers*; these blobbers would provide cheap storage, but with extremely low read rates. In their role, they could help NFTs to weather sudden, unexpected rises in the cost of storage relative to the value of ZCN.

Some client must initialize the transaction and pay the cost of that transaction. With 0Chain, certain types of transactions are designated zero-cost; the re-locking of NFT funding pools could be added to this list. Alternately, a portion of the minted tokens could be given to the client to compensate them for the transaction fee, or even to provide a small reward for calling the transaction.

The ability of the token-locking reward protocol to create a steady revenue stream seems likely to be useful in a number of other areas. Whenever an ongoing service needs to be funded, this design provides a model of how that funding could be achieved.

8   Conclusion and Future Work

In this paper, we have proposed a system for helping artists to produce NFTs on the blockchain. Our crowdsourcing mechanism both helps artists to create their new projects and more easily reward their backers with a share of the proceeds. We also show how the 0Chain blockchain could be leveraged to store large NFTs and how a revenue stream could be created to offset the cost of that storage.

In future work, we intend to explore how these NFTs could be transferred across blockchains. Additionally, we intend to expand our prototype to further explore the challenges of NFTs.

References

1. Real estate use cases for blockchain technology. Enterprise Ethereum Alliance - Real Estate Special Interest Group, vol. 1 (2019)
2. Austin, T.H.: SpartanGold: a blockchain for education, experimentation, and rapid prototyping. In: Park, Y., Jadav, D., Austin, T. (eds.) SVCC 2020. CCIS, vol. 1383, pp. 117–133. Springer, Cham (2021). https://doi.org/10.1007/978-3-030-72725-3_9
3. Chevet, S.: Land registry on blockchain. Blockchain Technology and Non-Fungible Tokens: Reshaping Value Chains in Creative Industries. Master's thesis, Paris, France (2018)

4. Cryptokitties, cryptopunks and the birth of a cottage industry. Financial Times (2018)
5. Cryptokitties key information. https://www.cryptokitties.co/technical-details. Accessed April 2021
6. Dash, A.: NFTs Weren't Supposed to End Like This. The Atlantic, Washington (2021)
7. Entriken, A.W., Shirley, D., Evans, J., Sachs, N.: EIP-721: ERC-721 non-fungible token standard (2018). https://eips.ethereum.org/EIPS/eip-721
8. Kim, M., Hilton, B., Burks, Z., Reyes, J.: Integrating blockchain, smart contract-tokens, and IoT to design a food traceability solution. In: 2018 IEEE 9th Annual Information Technology, Electronics and Mobile Communication Conference (IEM-CON), pp. 335–340 (2018). https://doi.org/10.1109/IEMCON.2018.8615007
9. Merkle, R.C.: Protocols for public key cryptosystems. In: 1980 IEEE Symposium on Security and Privacy, p. 122 (1980)
10. Merrill, P., Austin, T.H., Thakker, J., Park, Y., Rietz, J.: Lock and load: a model for free blockchain transactions through token locking. In: IEEE International Conference on Decentralized Applications and Infrastructures (DAPPCON). IEEE (2019)
11. Müller, M., Janczura, J.A., Ruppel, P.: DeCoCo: blockchain-based decentralized compensation of digital content purchases. In: 2nd Conference on Blockchain Research & Applications for Innovative Networks and Services, BRAINS 2020, Paris, France, 28–30 September 2020, pp. 152–159. IEEE (2020). https://doi.org/10.1109/BRAINS49436.2020.9223299
12. Nakamoto, S.: Bitcoin: a peer-to-peer electronic cash system (2008). https://bitcoin.org/bitcoin.pdf. Accessed 1 April 2021
13. Patil, M.: Land registry on blockchain. Master's thesis, San José State University, San Jose, CA, USA (2020)
14. Regner, F., Urbach, N., Schweizer, A.: NFTs in practice - non-fungible tokens as core component of a blockchain-based event ticketing application. In: Krcmar, H., Fedorowicz, J., Boh, W.F., Leimeister, J.M., Wattal, S. (eds.) Proceedings of the 40th International Conference on Information Systems, ICIS 2019, Munich, Germany, 15–18 December 2019. Association for Information Systems (2019). https://aisel.aisnet.org/icis2019/blockchain_fintech/blockchain_fintech/1
15. Reitwießner, C., Johnson, N., Vogelsteller, F., Baylina, J., Feldmeier, K., Entriken, W.: EIP-165: ERC-165 standard interface detection (2018). https://eips.ethereum.org/EIPS/eip-165
16. Salah, K., Nizamuddin, N., Jayaraman, R., Omar, M.: Blockchain-based soybean traceability in agricultural supply chain. IEEE Access 7, 73295–73305 (2019). https://doi.org/10.1109/ACCESS.2019.2918000
17. Selvi, S.S.D., Paul, A., Dirisala, S., Basu, S., Rangan, C.P.: Sharing of encrypted files in blockchain made simpler. In: Pardalos, P., Kotsireas, I., Guo, Y., Knottenbelt, W. (eds.) Mathematical Research for Blockchain Economy. SPBE, pp. 45–60. Springer, Cham (2020). https://doi.org/10.1007/978-3-030-37110-4_4
18. Stefanović, M., Ristić, S., Stefanović, D., Bojkić, M., Pržulj, D.: Possible applications of smart contracts in land administration. In: 2018 26th Telecommunications Forum (TELFOR), pp. 420–425 (2018). https://doi.org/10.1109/TELFOR.2018.8611872
19. Trautman, L.J.: Virtual art and non-fungible tokens (2021). https://papers.ssrn.com/sol3/papers.cfm?abstract_id=3814087
20. Vogelsteller, F., Buterin, V.: EIP-20: ERC-20 token standard (2015). https://eips.ethereum.org/EIPS/eip-20

21. Westerkamp, M., Victor, F., Küpper, A.: Blockchain-based supply chain traceability: token recipes model manufacturing processes. In: IEEE International Conference on Internet of Things (iThings) and IEEE Green Computing and Communications (GreenCom) and IEEE Cyber, Physical and Social Computing (CPSCom) and IEEE Smart Data (SmartData), iThings/GreenCom/CPSCom/SmartData 2018, Halifax, NS, Canada, 30 July–3 August 2018, pp. 1595–1602. IEEE (2018). https://doi.org/10.1109/Cybermatics_2018.2018.00267

22. Wood, G.: Ethereum: a secure decentralised generalised transaction ledger (2014). https://gavwood.com/paper.pdf

Automated Flag Detection
and Participant Performance Evaluation
for Pwnable CTF

Manikant Singh[✉] , Rohit Negi[✉] , and Sandeep K. Shukla[✉]

C3i Center, Department of Computer Science and Engineering,
Indian Institute of Technology, Kanpur, India
{manikant,rohit,sandeeps}@cse.iitk.ac.in

Abstract. The demand for cyber security awareness, education, evaluation of learning levels of students etc., has increased in the past few years. In order to meet this rising demand, several cyber security learning and training platforms have been developed. Capture the flag (CTF) platforms and cyber ranges have become primary tools that facilitate education, training and recruitment of cyber security personnel. These tools evaluate and rank the participants on the basis of challenges solved by them. A discrete evaluation mechanism focusing only on flags solved, fails to ensure that the effort and knowledge demonstrated by the participants while solving the challenge, are factored into the scoring system. Most of these tools do not even distinguish between participants actually solving the flags vs. those who might copy a captured flag without actually working on the problem. Further, in flag only scoring systems, participants feel discouraged as they fail to score without finding the flags – despite putting in enormous time and effort. In this paper, we present our novel approach to quantify participant's learning, efforts, and any unethical practices. We award partial scores by automatically capturing their behavior while solving the CTF problems. We also provide an accurate ranking system with automated solved challenge detection which replaces the need for manual flag submission. In our system, participants get hybrid scores based on their efforts, and organizations get a better and an effective evaluation tool.

Keywords: Cyber security · CTF · Binary exploitation · Reverse engineering

1 Introduction

Cyber security education and training play an essential role in safeguarding organizations and their users from cyber attacks. It helps in making people aware of the extent and the severity of cyber threats and encourages them to improve their security posture. Several cyber security courses and hands-on exercises have been designed to facilitate education, training & recruitment of cyber security professionals. Hands-on exercises using *capture the flag* (CTF) competitions & cyber

ⓒ The Author(s) 2022
S.-Y. Chang et al. (Eds.): SVCC 2021, CCIS 1536, pp. 126–142, 2022.
https://doi.org/10.1007/978-3-030-96057-5_9

ranges are considered more effective as they are engaging and they acquaint the participants with core concepts in short time-frames. Consequently, experienced instructors have shifted to using hands-on learning [9] via CTFs. CTFs have been used in the cyber security domain for education as well as evaluation of learning, for over a decade [3,5,6,8].

In standard CTFs, participants apply theoretical concepts to solve/exploit the challenges and capture the hidden flags which are only visible once a successful exploit of one or multiple vulnerabilities has been made.

In most CTF platforms, the participants have to submit the flags on an online portal to be awarded scores for it. If they fail to submit a flag, they get no score at all even if they spent a lot of effort unsuccessfully and almost solved it. Administering organization of the CTF use the obtained scores to evaluate, rank, select or measure learning outcomes.

Stakeholders of the CTF may differ based on their use cases. For educational use cases, the students and the instructors will be the stakeholders. In case a student obtain no score even after putting in a lot of effort, he/she might lose interest in the curriculum. On the other hand, the instructors may fail to evaluate the overall progress of their students. For a recruitment use case, the examiner, the examinees, and the hiring organization are the stakeholders. With the discrete scoring (i.e. based only on the submission of flags), hiring organizations may not get potential candidates.

Most existing tools and frameworks use discrete evaluation scheme. Most of the existing tools do not even support automated detection of capturing of flags. The first challenge is that participants in the CTF are being graded on the basis of submission of flag. Second is, they are ranked on the basis of the timestamps of the submission of flags and not on the time stamps of flag capture events. The third problem is unethical practices by the participants. It is therefore essential to have a *cyber security education and recruitment framework* (**CSERF**) that aims to fill these gaps by systematic evaluation of participants. An ideal CSERF must facilitate not only training exercises but also capture trainee's activities, automatically detect solved challenges, and penalize those making unethical attempts – including attack on other participants or compromising the platform itself.

In this work, we introduce a novel method to capture a participant's pursuits and quantify their efforts. This paper presents a hybrid scoring scheme where partial scores are awarded to the participants for their struggle/effort in solving the problems. Partial scores are awarded, provided that they attempt a question. Our ranking system is based on the correct sequence of events leading to solving the problem, and on penalizing detected attempts at unfair means to problem solving. Moreover, automatic detection of solved problem removes the need to create any distinct flag submission mechanism for the participants.

Rest of the paper is organized in the following manner: Sect. 2 is a brief discussion on related work. Sect. 3 describes our setup and testing environment. Sect. 4 details our evaluation scheme. Section 5 shows the results of a case study. Section 6 concludes the paper with some indications of future plans.

2 Background and Related Work

Capture the flag is one of the prominent & dominating setup used in cyber security education and evaluation. Jeopardy style CTF is the most popular type of event. As per Valdemar Švábenský et al. [11], it constitutes 86% of all the CTFs listed on ctftime.org and the challenges are further subdivided into multiple categories as mentioned in Table 1.

Table 1. CTF categories and vulnerabilities tested

Category	Vulnerabilities
Binary exploitation	Buffer/Integer overflow, format string, return oriented programming, heap exploitation, double free, use after free, etc.
Web exploitation	Broken Access Control, Broken Anti-Automation, Broken Authentication, Cross Site Scripting (XSS), Cryptographic Issues, Improper Input Validation, Injection, Insecure Deserialization, Security Misconfiguration, Security through Obscurity, Sensitive Data Exposure, Unvalidated Redirects, Vulnerable Components, XML External Entities (XXE), etc.
Network exploitation	DNS Spoofing, DNS poisoning, MITM (Man-in-the-middle), Packet Sniffing, ARP spoofing, IP spoofing, timing attack, session hijacking, etc.
Crypto	Brute force, replay, hash collision, side channel attack, etc.

Table 2. Feature comparison between different software

Features	CSERF	Pwnable.kr	CyTrOne	Hack The Box	The Juice Shop
Educational use	√	√	√	√	√
Score board	√	√	√	√	√
Behaviour analysis	√	X	X	X	X
Automated flag detection	√	X	X	X	√
Automated scoring	√	X	X	X	√
Partial scoring	√	X	X	X	X

In Table 2, we have compared several CTF tools that we used and studied by comparing their features against desired features. CTFd [4] is an open-source platform that helps to host CTF with user and flag score management capabilities. In the past few years, it has become a de-facto standard to host capture the flag tournaments via CTFd. It lets participants to submit flags for each problem they solve and unlock new challenges as they progress. CyTrOne [1] was developed by the Cyber Range Organization and Design (CROND) at the Japan Advanced Institute of Science and Technology (JAIST). It is a cyber security training framework that simplifies the training setup process by integrating training content and training environment management. Open Cyber Challenge Platform (OCCP) is an open-source platform. It is used to educate and train high school/college students about cyber attacks by recreating already discovered attacks in a controlled environment. Participants defend/attack/investigate a network and data center with realistic attacks, but the community no longer supports it. Pwnable.kr [3] is a non-commercial war game website specialised in hosting pwnable CTF. It's similar to OverTheWire website but with some additional graphics to make it more engaging for the end-user. Another online platform is HackTheBox which allows participants to test their penetration testing skills and have a global scoreboard for teams worldwide. Multiple companies have used it for selecting new hires. All these platforms/frameworks use the conventional method of finding and submitting the flag to score.

OWASP TheJuiceShop is one of the best platforms to learn about web security. It automatically maintains a scoreboard based on the sequence of events. Once the participant accesses a particular file or performs a similar attack, it generates an event and automatically detects solved challenges. Again a discrete evaluation scheme is followed and has no partial scoring feature based on participant efforts in the right direction.

For a fine-grained evaluation, S.K. Kim et al. [7] proposed a platform to provide a fine-grained evaluation technique for pwnable CTF. They have defined four levels to evaluate student's progress: 1) Crash Check, 2) Control Flow Handling Check, 3) Mitigation Bypassing Check, and 4) Full Exploit Check. These steps are associated with the level of understanding a student might be possessing if they successfully pass a particular evaluation step. Steps are arranged in increasing order of difficulty. If a student fails at a particular level, the student is considered to have knowledge of the techniques required to cross all previous levels. Though the framework is designed to mitigate an instructor's burden, they still need to create build scripts and select the types of mitigation through a web portal. In this paper, we aim to fill the following gaps of existing CTF platforms. Most platforms provide marks solely based on flag submission, and no marks are given for efforts or evidence of knowledge and learning. Furthermore, they do not capture candidates participation behaviour data and do not apply any behavioural analytics on captured behaviour. Existing platforms do not distinguish between a flag obtained using unfair means versus flags obtained through genuine efforts.

In our work, we capture participant's activities and do behavioural analysis to generate a hybrid score. The automated multidimensional grading would enhance the capability of CTF platforms/tools and generate realistic scores and reduce the time to assess the skill set of participants. In addition, an automated flag detection system provides more accurate rankings based on the precise time stamps of flag retrieval and reduces the possibility of using unfair means (such as copy another participant's flags and submitting them).

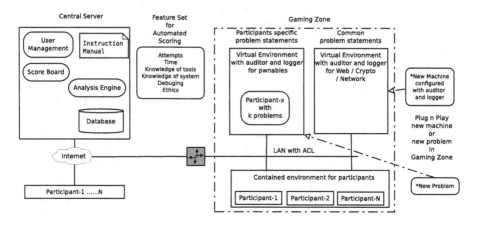

Fig. 1. Platform architecture

3   Proposed Framework

This section explains our platform architecture, system setup and techniques used to extract grading parameters and grading scheme.

3.1   High Level System Architecture

As depicted in Fig. 1, high level platform architecture has two major components. The first component is the *game zone*, where all the participant machines having problem setups are deployed with auditor and logger. The second component is the *central server*, where all the logs pertaining to a participant's activity will be processed for automated scoring. This system configuration ensures that the vulnerable machine is isolated from the Internet and cannot be used to attack other systems present on the Internet.

A single virtual machine per participant, is provisioned for the entire duration of a CTF and all the participants are given access to their separate set of vulnerable binaries to exploit. For any new challenge, we only have to add that challenge to the provisioned machines.

The Gaming Zone. It is a collection of isolated controlled virtual environments with pre-configured auditors and logging tools that record participant's behaviour. New problems can be added to the existing virtual environment at any point in time. New virtual machines can be added to the gaming zone at the trainer/instructor's discretion.

Centralised Server. The Centralized server is responsible for user and scoreboard management. It also consists of an Analysis engine, which is the core of the framework. Figure 2 shows different stages of analysis engine and the sequence of operations to auto-grade participants.

The Analysis engine can be considered as a collection of three sub-modules 1) Data collector 2) Analyzer 3) Auto-Grader. The first and second modules are responsible for gathering and extracting raw information. The third module utilises collected information to generate scores. The responsibilities of the first module are further divided into the following steps.

A) Update Participants List: Before retrieving the data from the server, the agent checks if there is any addition or removal of users from the server and update the local list of users. All other operations will be performed only for users present in the updated list of users.

B) Get Latest Data Logs: Once the users list is updated, the agent fetches the most recent logs generated by Osquery [10], GDB history and shell history.

C) Parse Unprocessed Logs (Analyzer): At this stage, the engine processes all the recent logs and extract the relevant information in JSON format. Later these parsed logs are utilised by the auto-grader for partial grading.

3.2   Behavioral Analysis and Data Sources

As no data set was readily available for our research purpose, we conducted a capture the flag tournament in a cyber security training course at our institute, consisting of sixty-nine participants and recorded their live data. More details of the tournament are specified in Sect. 5.

This section explains the techniques used to capture a user's behaviour and the source of each data point. The Data sources can be classified into the following four categories.

Command Line Process. We track all the processes invoked by a user either from a terminal or by a process started from the terminal. All processes invoked directly or indirectly from the terminal are labelled as command-line processes. We keep a check on the execution of all the executable files with the help of Osquery.

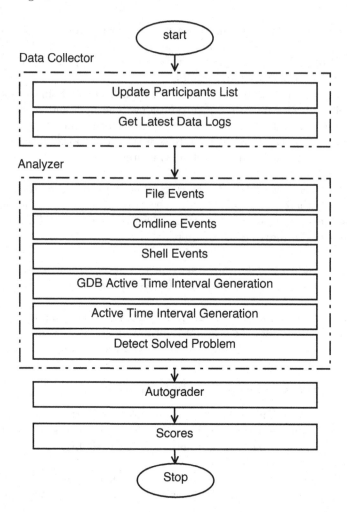

Fig. 2. Analysis engine

File Events. In addition to terminal driven processes, we also keep track of files accessed by a participant. Rules are set to watch for any access made to files. Keeping track of the time of access helps us determine the liveness of the participant on the server. The file path which is accessed helps us determine the binary or the file on which participant is currently working.

GDB History. GDB is an essential tool for the process of binary exploitation. Consequently, we also collect the GDB history of each user. However, by default, GDB only keeps track of commands entered by a participant, which is not sufficient for our use case. We need to know at which executable the user initiated the GDB session and how long the session lasts.

To achieve this, we have modified the GNU Debugger source code to record the executable name on which GDB is invoked and the timestamp of each command entered during that session. Using this technique, we accurately determine the debugging activities of users.

Shell History. Shell history is collected from the history of each user. This lets us determine all the activities performed by the user from the terminal.

Other data sources such as open sockets and listening ports were found irrelevant during the feature selection process. Therefore they are not discussed here.

4 Our Approach

4.1 Feature Set

In this section, we elucidate each feature extracted by our analysis engine that contributes to the final grading.

Number of Attempts: To inspect whether a participant attempted a problem or not, we look at the command line processes and the shell history. If the occurrence of a challenge's executable file name is missing, we say that the participant didn't attempt the question. If the occurrence is found, we keep track of its frequency. Here C is the set of all challenges, and A_{pc} denotes the number of attempts made to solve a challenge $c \in C$ by the participant $p \in P$ (Table 3). The maximum value of attempts made to solve a challenge c among all participants is given as

$$A_c^{max} = \max_{p \in P} (A_{pc}) \tag{1}$$

Table 3. Feature set with respective weight

Feature	Symbol	Weight
No. of attempts	A	W_A
Active time intervals	I	W_I
GDB active time	G	W_G
Knowledge of tools	K^t	W_{K^t}
Knowledge of system	K^s	W_{K^s}
Ethics	E	W_E

Active Time: Linux provides built-in tools to determine when the user logs in and how long the log-in session lasts. However, this gives us an overall active time but to correctly estimate the time spent by each participant on a particular problem, we propose an algorithm to generate activity periods from the logs synthetically. We have defined a parameter called buffer time (BT), which is the maximum time difference (in seconds) between two consecutive activities of the same session.

initialization;
BT – Buffer Time
I – Superset of active intervals, initially empty
P – Set of participants
C – Set of challenges
for $p \in P$ **do**
    for $c \in C$ **do**
        $I_{pc} \leftarrow 0$ /* Sum of active intervals for a challenge c of the participant p */
        T_{pc} – Timestamps of activities for a challenge c of the participant p
        $N \leftarrow size(T_{pc}) - 1$;
        $T_{pc} \leftarrow sorted(T_{pc})$;
        $S_{pc} \leftarrow T_{pc}^0$; /* Active interval start time */
        $E_{pc} \leftarrow T_{pc}^0$; /* Active interval end time */
        $AI_{pc} \leftarrow (E_{pc} - S_{pc})$; /* Current active interval */

        for $T_{pc}^i \in \{T_{pc}^1 \dots T_{pc}^N\}$ **do**
            if $T_{pc}^i - T_{pc}^{i-1} > BT$ **then**

                /* End previous active interval */
                $AI_{pc} \leftarrow (T_{pc}^{i-1} - S_{pc})$;
                /* Add active interval AI_{pc} to I_{pc} */
                $I_{pc} \leftarrow AI_{pc} + I_{pc}$;
                /* Update start and end time */
                $S_{pc} \leftarrow T_{pc}^i$;
                $E_{pc} \leftarrow T_{pc}^i$;
             end
            else
                /* Extend active interval AI_{pc} end time*/
                $E_{pc} \leftarrow T_{pc}^i$;
             end
        end
        /* End last open active interval */
        $AI_{pc} \leftarrow (E_{pc} - S_{pc})$;
        $I_{pc} \leftarrow AI_{pc} + I_{pc}$;
        /* Insert sum of active interval I_{pc} into I */
        $I \leftarrow I_{pc} \cup I$;
    end
end

Algorithm 1: Active Interval Generation

Algorithm 1 shows a procedure for determining accurate active time intervals. Let us say that a participant is writing a payload script in the challenge's directory. With the help of an auditing tool, we watch for changes in a directory/file and record timestamps for each activity performed on a particular file. Using these timestamps, we generate a synthetic active time interval.

Let us say that a participant p is working on challenge c. While working on the challenge, the participant performs following actions.

- Creates a script payload.py at (T_{pc}^i)
- Modifies payload.py at (T_{pc}^{i+1})
- Deletes palyload.py at (T_{pc}^{i+2})

If the time difference between these consecutive activities is less than the buffer time (BT) and there is no other activity in-between, we say that the participant started working on challenge c at T_{pc}^i and continued till T_{pc}^{i+2}. Therefore, they contribute to the same active interval AI_{pc}. If the time difference between two consecutive activities is greater than the buffer time (BT), we conclude the active time interval, add it to sum of existing intervals I_{pc} and begin new active interval for the current activity at time T_{pc}^i, where I_{pc} is sum of all such active intervals for participant p and challenge c. Here I_{pc} is an integer. The maximum value of cumulative interval for a challenge c among all participants is given as

$$I_c^{max} = \max_{p \in P} (I_{pc}).$$
(2)

This approach gives us a reasonably good estimate of a user's cumulative active time for a particular challenge.

GDB Sessions and Active Time: With our modified GDB, we have executable names on which the GDB was invoked and the timestamp of each command entered. We apply the same Algorithm 1 for synthetic timestamp sequencing on GDB activities as well, so that we have accurate active time estimates and the user is not able to manipulate active sessions. G_{pc} is a sum of all GDB active intervals for a participant p and challenge c.

The maximum value of cumulative GDB session interval for challenge c among all users is given as

$$G_c^{max} = \max_{p \in P} (G_{pc}).$$
(3)

Knowledge of Tools: To check if a user has used the tools required to crack the problem, we track usage of tools like readelf, objdump, ltrace, strace, ptrace, hexdump, nasm, gdb and file. This takes care of assessing the granularity of knowledge. The list can also be modified before starting the analysis engine. K_p^t denotes the cumulative count of selected tools used by a participant p. The maximum value of K_p^t among all participants is given as

$$K_{max}^t = \max_{p \in P} (K_p^t).$$
(4)

Knowledge of System: Similar to tools knowledge, we also check whether a participant understands Linux and its permission/privilege system. To take care of this parameter, we have identified few commands like sudo, chmod, chroot, chgrp and chown, which the participants use during an actual CTF session. However, repetitive usage of such commands indicates that the user does not understand Linux and its permission system. Commands to watch for this purpose can be added/removed at the trainer's discretion. K_p^s denotes the cumulative count of selected commands used by participant p. The maximum value of K_p^s among all participants is given as

$$K_{max}^s = \max_{p \in P} \left(K_p^s \right). \tag{5}$$

Ethics: To identify unethical attempts, we check if participants try to access restricted inodes. Though in most pwnable/binary CTFs, participants cannot view other users' directory content, we track such events so that the trainer gets an idea about participants trying to use unfair means. For example, commands like

"cat ../../../home/otheruser/problem/file.txt"
"ls ../../../home/otheruser/problem/"
"ls /home/otheruser/problem"

all are considered unethical attempts, E_p denotes the frequency of such attempts made by participant p. Participants are penalized for these activities. The maximum value of E_p among all participants is given as

$$E_{max} = \max_{p \in P} \left(E_p \right). \tag{6}$$

Solved Challenge Auto Detection: Our framework automatically detects when a participant successfully opens a "flag.txt" with elevated privileges and mark that challenge as solved for that particular participant. The system also maintains the time of the actual flag retrieval to ensure a precise ranking system. Our technique is independent of challenge type (buffer overflow, integer overflow etc.) and flag value written in flag.txt, making it easier for the tutor to have randomized flag value for each participant. Note that the auto detection mechanism for other kinds of challenges such as web security challenges will be different.

initialization;
/* $\{A, I, K^t, K^s, G, E\}$ are features with $\{W_A, W_I, W_{K^t}, W_{K^s}, W_G, W_E\}$ as their respective weights.*/
/* Set of dedicated marks for challenges*/
$CM \leftarrow \{CM_1, CM_2 \ldots CM_m\};$
F – Analysis time frame in seconds
TS_D – Timestamp indicating start of the current day
SF – /* Scaling factor */
$CS_p \leftarrow 0;$ /* Cumulative score of $p \in P$ */
$CE_p \leftarrow 0;$ /* Cumulative effort score of $p \in P$ */
$M_p \leftarrow 0;$ /* Marks to maintain rank of $p \in P$ */
$HS_p \leftarrow 0;$ /* Hybrid score of $p \in P$ */
for $p \in P$ **do**
> **for** $c \in C$ **do**
>> /*normalize data */
>>
>> $\overline{I}_{pc} \leftarrow \frac{I_{pc}}{I_c^{max}};$
>>
>> $\overline{G}_{pc} \leftarrow \frac{G_{pc}}{G_c^{max}};$
>>
>> $\overline{A}_{pc} \leftarrow \frac{A_{pc}}{A_c^{max}};$
>>
>> **if** c *is solved* **then**
>>> t – Initial timestamp when c was solved
>>> **if** $t >= TS_D$ **then**
>>>
>>> **else**
>>>> /*Add dedicated challenge score and incentivize early solver */
>>>> $M_p \leftarrow M_p + CM_c + SF * \left(1 - \frac{t - TS_D}{F}\right);$
>>>
>>> **end**
>>> $M_p \leftarrow M_p + SF;$
>>
>> **else**
>>> $CS_p \leftarrow CS_p + \frac{W_A * \overline{A}_{pc} + W_I * \overline{I}_{pc} + W_G * \overline{G}_{pc}}{W_A + W_I + W_G};$
>>
>> **end**
>
> **end**
> /*normalize data */
>
> $\overline{K^t} \leftarrow \frac{K_p^t}{K_{max}^t};$
>
> $\overline{K^s} \leftarrow \frac{K_p^s}{K_{max}^s};$
>
> $\overline{E} \leftarrow \frac{E_p}{E_{max}};$
> /* Weighted score of knowledge and ethics */
>
> $CE_p \leftarrow \left(\frac{W_{K^t} * \overline{K^t} - W_{K^s} * \overline{K^s} - W_E * \overline{E}}{W_{K^t} + W_{K^s} + W_E}\right) + CS_p;$

end
$CE^{max} \leftarrow \max_{p \in P}(CE_p);$
for $p \in P$ **do**
> /* Normalize cumulative effort */
>
> $CE_p \leftarrow \left(SF * \frac{CE_p}{CE^{max}}\right);$
> /* Update hybrid score of participant p */
> $HS_p \leftarrow CE_p + M_p;$

end

Algorithm 2: Evaluation Algorithm

4.2 The Grading Algorithm

In the previous section, we discussed all the features which contribute to the final evaluation. Since every tutor may not find all these parameters equally relevant to them, we let them define the weights $\{W_A, W_I, W_{K^t}, W_{K^s}, W_G, W_E\}$ for each parameter. Given the weights, the auto-grader comes into action to calculate scores for all the participants. The Algorithm 2 utilizes data points $\{I_{pc}, G_{pc}, A_{pc}, K_p^t, K_p^s, E_p\}$ collected at the behavioural analysis stage to calculate these scores.

In our Algorithm 2, we loop on every participant p, and on each challenge $c \in C$. For a particular participant p, we first check if a challenge is solved or not. Our system automatically detects the first instance when the challenge was solved and imparts the dedicated challenge score CM_c and an additional incentive score to an early solver. Incentive score is calculated on the basis of the initial timestamp when a challenge was solved, as shown in the Algorithm 2. If the challenge was solved at an earlier time frame, we impart the maximum incentive marks that can be given for a time frame. The Incentive score ranges between 0 to 1 and scaled with scaling factor SF as per the configuration. Sum of CM_c and incentive score contributes to M_p, that is, the score to maintain the rank for participant p. The incentive score helps a participant maintain his/her prior rank similar to any other standard flag submission based ranking system.

If a challenge is not solved, then we calculate the partial score from the weighted sum of parameters $\{I_{pc}, G_{pc}, A_{pc}\}$ which are exclusively related to a challenge c and participant p. This weighted sum is stored in a temporary variable called cumulative score CS_p. Other parameters $\{K_p^t, K_p^s, E_p\}$ contributes to partial scores for a participant p, and the weighted score of these parameters combined with CS_p gives us the cumulative effort score CE_p for a participant p for a given analysis time frame.

Finally, hybrid score HS_p, a combination of rank maintainer score M_p and cumulative effort score CE_p, is given to the participant p. The final ranking is done using this hybrid score.

5 Case Study

Since there was no existing data set for our research and analysis, we hosted a capture the flag (CTF) tournament for a cyber security training course we run at our institute, with a total of sixty-six participants and collected logs for their activities while they were exploiting the given challenges in our controlled environment. Each participant had access to the gaming zone with pre-configured auditors and loggers. All participants actively tried to crack the given challenges, only 18.18% participants were able to solve at least one challenge, and 43.93% participants were found to have issues with tools and system.

In addition to our automated flag detection system, we gave them the facility to submit flags manually on a ranking server (CTFd) for live tracking.

Table 4 shows the ranking obtained by top 12 participants based on their flag submission and Table 5 shows the ranking obtained with our framework.

Table 4. CTFD ranking system based on flag submission

Rank	Name	Score	Last submission
1	P_{35}	3600	2021-05-08 23:51:07
2	P_{34}	3600	2021-05-20 09:26:39
3	P_5	2600	2021-05-20 02:11:09
4	P_{61}	2600	2021-05-29 17:26:43
5	P_8	2600	2021-05-29 19:14:19
6	P_{43}	2100	2021-04-27 20:36:54
7	P_{55}	1900	2021-05-16 22:47:22
8	P_{52}	1400	2021-05-30 12:26:20
9	P_{23}	1300	2021-05-28 12:56:27
10	P_{42}	800	2021-04-15 00:19:07
11	P_{12}	600	2021-04-19 13:19:36
12	P_{64}	600	2021-05-28 16:45:55

Table 5. Ranking with automated flag detection

Rank	Name	Score	Last solve
1	P_{35}	3600	2021-05-08 23:49:28
2	P_{34}	3600	2021-05-20 09:25:51
3	P_5	2600	2021-05-20 02:10:00
4	P_{61}	2600	2021-05-29 17:22:55
5	P_8	2600	2021-05-29 19:13:05
6	P_{43}	2100	2021-04-27 20:36:54
7	P_{55}	1900	2021-05-16 22:47:06
8	P_{52}	1400	2021-05-30 00:22:48
9	P_{23}	1300	2021-05-28 00:55:26
10	P_{42}	800	2021-04-15 00:15:51
11	P_{12}	600	2021-04-17 19:46:38
12	P_{64}	600	2021-05-28 16:45:25

Participants with zero submission got ranking based on hybrid scores for their efforts. This technique let us identify participants who gave significant efforts yet failed to solve. Table 7 shows top 5 participants who solved zero challenges yet got ranking from 13 to 17 based on their pursuit.

From Tables 4 and 5, we can see that there is always a delay of few minutes between flag retrieval and flag submission. Difference between flag retrieval and submission is significant in case of participant P_{12} ranked at position 11. These ranks could have been different if some other participant submitted their flag in between. Based on these results, we can conclude that ranking based on flag

Table 6. Ranking combined with hybrid scores

Rank	Name	Original score	Cumulative effort score (CE)	Hybrid score (HS = CE + M)
1	P_{35}	3600	2132.50	5732.50
2	P_{34}	3600	1359.36	4959.36
3	P_5	2600	1221.96	3821.96
4	P_8	2600	853.03	3453.03
5	P_{61}	2600	802.71	3402.71
6	P_{43}	2100	1294.26	3394.26
7	P_{55}	1900	888.81	2788.81
8	P_{52}	1400	244.15	1644.15
9	P_{23}	1300	150.30	1450.30
10	P_{42}	800	468.06	1268.06
11	P_{12}	600	476.32	1076.32
12	P_{64}	600	83.86	683.86

Table 7. Top 5 participants with zero solves

Rank	Name	Original score	Hybrid score (HS = CE + M)
13	P_7	0	161.99
14	P_4	0	109.14
15	P_{48}	0	88.50
16	P_2	0	50.48
17	P_{13}	0	48.11

submission does not fully capture and rank the participants based on the criteria we believe should be used in evaluating participants. However, our framework more accurately capture these criteria in scoring and ranking of participants. Of course, one case study is not enough to be sure, and we plan to do more extensive experiments in the future batches of trainees to get a better data support for this assertion.

Table 6 shows ranking obtained after assigning partial scores (cumulative effort scores). Here we can clearly see that partial scores works as a tie breaker and favours participants with more efforts in lesser time frame. The justification for this is that those putting more effort in solving the challenges are possibly learning more from the exercises.

6   Conclusion and Future Work

CTF tournaments are playing a crucial role in cyber-security education and training activities. We have built a system that improves the evaluation techniques used in traditional CTF tournaments, and this paper tries to elucidate our work. Our framework's evaluation scheme motivates participants to put in more efforts and learn, through partial scores for their participation and activeness. They are not judged just on the basis of solved problems but also ethics, and their knowledge of tools and systems. From Tables 5 and 6, we can observe that with hybrid scores, we still get similar ranks as that of a discrete score based ranking system but with some significant positive discrimination for hardworking and ethical participants. From Table 7, we observe that participants who solved zero challenges also got their ranks aligned according to their efforts, which does not happen in a discrete score-based ranking system. Discrete score based systems do not provide any way to differentiate between active and non-active participants. Techniques proposed in this paper creates a favourable environment for cyber security organizations, educational institutions and participants. With solved challenge auto-detection, we have improved the ranking system based on correct sequence of events.

Although CSERF is the first of its kind to auto-grade participants, there is still a vast scope of improvement and innovation. The Framework is implemented for pwnable CTF. Nonetheless, there are numerous training challenges like networking, web, cryptography, defensive cyber security etc., where we can extend the support for hybrid scoring. Harsh et al. [2] mentioned that using server logs, one can identify participant's behaviour. Web request can be classified into attacks such as SQL Injection, Cross-Site Scripting, Path-traversal, Command Injection, Cross-site request forgery etc. We can utilize this technique to calculate the partial score of participants for web challenges also.

Also, as we gather more data over many tournaments, scoring system can be improved using machine learning models. Currently, the data being limited, we have provided algorithms based on our assumptions on how a user behaves towards solving CTF challenges.

References

1. Beuran, R., Pham, C., Tang, D., Chinen, K., Tan, Y., Shinoda, Y.: CyTrONE: an integrated cybersecurity training framework (2017)
2. Bhagwani, H., Negi, R., Dutta, A.K., Handa, A., Kumar, N., Shukla, S.K.: Automated classification of web-application attacks for intrusion detection. In: Bhasin, S., Mendelson, A., Nandi, M. (eds.) SPACE 2019. LNCS, vol. 11947, pp. 123–141. Springer, Cham (2019). https://doi.org/10.1007/978-3-030-35869-3_10
3. Azam, M.H.B.N., Beuran, R.: Usability evaluation of open source and online capture the flag platforms (2018)
4. Chung, K.: Live lesson: lowering the barriers to capture the flag administration and participation. In: 2017 USENIX Workshop on Advances in Security Education (ASE 2017) (2017)
5. Chung, K., Cohen, J.: Learning obstacles in the capture the flag model. In: 2014 USENIX Summit on Gaming, Games, and Gamification in Security Education (3GSE 14) (2014)

6. Ford, V., Siraj, A., Haynes, A., Brown, E.: Capture the flag unplugged: an offline cyber competition. In: Proceedings of the 2017 ACM SIGCSE Technical Symposium on Computer Science Education, pp. 225–230 (2017)
7. Kim, S.-K., Jang, E.-T., Park, K.-W.: Toward a fine-grained evaluation of the Pwnable CTF. In: You, I. (ed.) WISA 2020. LNCS, vol. 12583, pp. 179–190. Springer, Cham (2020). https://doi.org/10.1007/978-3-030-65299-9_14
8. Kucek, S., Leitner, M.: An empirical survey of functions and configurations of open-source capture the flag (CTF) environments. J. Netw. Comput. Appl. **151**, 102470 (2020)
9. McDaniel, l., Talvi, E., Hay, B.: Capture the flag as cyber security introduction. In: 2016 49th Hawaii International Conference on System Sciences (HICSS), pp. 5479–5486. IEEE (2016)
10. Osquery. Osquery
11. Švábenský, V., Čeleda, P., Vykopal, J., Brišáková, S.: Cybersecurity knowledge and skills taught in capture the flag challenges. Comput. Secur. **102**, 102154 (2021)

Towards Securing Availability in 5G: Analyzing the Injection Attack Impact on Core Network

Manohar Raavi[1]([✉]) [iD], Simeon Wuthier[1] [iD], Arijet Sarker[1] [iD], Jinoh Kim[2] [iD], Jong-Hyun Kim[3] [iD], and Sang-Yoon Chang[1] [iD]

[1] University of Colorado, Colorado Springs, CO 80918, USA
{mraavi,swuthier,asarker,schang2}@uccs.edu
[2] Texas A & M University, Commerce, TX 75428, USA
jinoh.kim@tamuc.edu
[3] Electronics and Telecommunications Research Institute, Daejeon, South Korea
jhk@etri.re.kr

Abstract. 5G technology for mobile networking involves control communications to set up the radio channels and the authentication and security credentials. The control communications preceding the authentication and subscription verification remain vulnerable against the communication injection threats. We study the injection threats on control communications in 5G New Radio standard in 3GPP. From our 5G client-based implementation and experimentation against real-world networking, we analyze and measure the threat impact against the 5G service provider infrastructure of the core network. To secure 5G networking, our paper discusses about future research directions for increasing the understanding of such vulnerability/threat and for building greater security and availability for 5G networking against such wireless injection threats.

Keywords: 5G · Mobile networking · Communication · Injection attack · DoS

1 Introduction

Mobile users use cellular networking to access the Internet and network with other computers. Since the 2G supporting text messaging in the early 1990's, cellular technologies have been evolving with the subsequent generation of technologies with increasing performances in communication rate, scalability/density in the number of users, and broader set of applications. This paper focuses on the most recent cellular technology in 5G networking.

In computer networking, the user devices use the *control communications* to set up the communication channels before delivering the goodput data using the channels. The control communications include the synchronization and the agreement in both the channel resources (such as the frequency band/bandwidth

© The Author(s) 2022
S.-Y. Chang et al. (Eds.): SVCC 2021, CCIS 1536, pp. 143–154, 2022.
https://doi.org/10.1007/978-3-030-96057-5_10

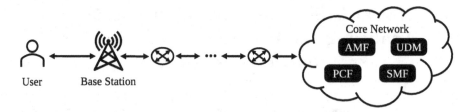

Fig. 1. The networking illustration including user, base station, and the core network. The communications between the user and base station are wireless (the furthest left arrow in the diagram), while the rest of the communication links are wired (the other arrows). Our injection threat affects all the communication links between the user and the core network.

and the modulation type), the security setup and credentials (such as the security mechanisms/protocol type and the key), and the application-layer registration. As shown in Fig. 1, in 5G and earlier cellular technologies, the control communications begin with the user initiating the channel request and the wireless channel setup, which involves the user and base station. Afterward, the control communication involves the core network beyond the base station for registration processing/verification and the security setup/key exchange. Therefore, the earlier the control communications before the security setup and authentication and the corresponding protection remain vulnerable against injection attacks.

In this paper, we study the security threat based on injections in the 5G control communications against the 5G networking service provider's availability. We show that the injection attack is feasible in the current 5G networking and quantify how much resources it consumes against the networking availability. While multiple injections lead to flooding for DoS attack, we focus on per injection and the corresponding resource consumptions at both the attacker (client) and the victim (the service provider end). To analyze and estimate the injection threat impact, we experiment on the real-world 5G networking to provide reference measurements and estimate the injection attack impact based on those measurement results.

The rest of the paper is organized as follows. Section 2 provides the background information on 5G control communications and Sect. 3 describes the injection threat against 5G. Section 4 provides the experiment methodology used, discusses measurement results, and estimates the injection impact. We review the related work on control communications security in Sect. 5 and discuss the future scope and potential countermeasures in Sect. 6. Finally, we conclude our work in Sect. 7.

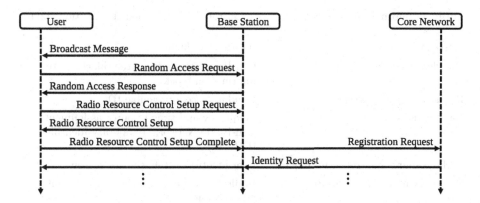

Fig. 2. 5G control communications including the wireless/radio access setup and the user registration verification [2,4]. This diagram focuses on the parts of the communication most relevant to our work, which are the parts before establishing the security and key exchange. The control communications further encapsulate the rest of the registration with "...", including authentication and security mode setup. The data communication for the actual 5G service/goodput follows after control communications.

2   Background of 5G Control Communications

In 5G, the user needs to establish control communications with the base station and core network to access the network services [2,4]. The control communications build on the a priori registration, which occurs before the real-time control communications described in Fig. 2; in such a priori registration, the user obtains a subscriber identity module (SIM) and its SIM registration gets inputted to the Core Network system. The real-time control communications which occur when the user wants to start networking for cellular service access verifies the user identity/subscription within the core network.

As shown in Fig. 2, these control communications involve performing channel setup with the base station and sending a registration request to the core network. Each of the individual control messages contains different parameters as listed in Table 1. The user synchronizes the downlink information from the broadcast messages (or synchronization signals) of the base station and starts a random access procedure for uplink connection during channel setup. There are 64 possible preambles defined in [3] and user equipment receives this information from broadcast messages.

For random access procedure initialization, the user sends a random access request to the base station with a preamble (randomly selected from the 64 preambles). In contention-free random access, the base station assigns a preamble. The base station prepares a random access response with allocated identifiers and provides resources to the user for further communications after the reception of the random access request from the user. The random access response is notified to the user through downlink control channel and carried on downlink shared channel.

Table 1. Information included in the 5G control communications transactions [2,4].

Message	Information includes
Broadcast message	Cell ID, master information block, system information block
Random access request	Random access preamble ID
Random access response	Frequency and time resource assignment, Cell-Radio Network Temporary Identifier (CRNTI)
Radio resource control setup request	User identity, Establishment reason
Radio resource control setup	Radio bearers configuration, master cell information
Radio resource control setup complete	Acknowledgement
Registration request	Registration type, 5G-globally unique temporary identifier, user capabilities
Identity request	Identity request message identity, identity type

The user, on receiving the random access response, sends a radio resource control connection setup request to the base station with a random number (ranges from 1 to 2^{39}-1) for the user identity (called User-Equipment Identity in the 3GPP standard [2,4]) and establishment reason. The response (radio resource control setup) from the base station contains radio bearers configuration and master cell information. Once the user receives the information, it configures the connection and sends a radio resource control connection complete message along with a registration request to the base station.

The base station forwards the registration request of the user to the core network. The core network has different logical functions to perform identity-check, authentication of the user, security setup, and user capability inquiry before accepting the registration. On successful registration, core network schedules the resources required to provide service to the user. Few of the core network functions are listed in Table 2 with the terminology used in our paper and their equivalents in 3rd Generation Partnership Project (3GPP) standards.

3   Injection Threat Against 5G

In this section, we describe and explain the threat model, the injection threat mechanism, and the vulnerability enabling the injection.

Threat Model. We assume a standard active wireless attacker model in that the attacker has the *injection* capability for generating and transmitting the wireless packets in 5G (which requires the radio/antenna frontend hardware and the wireless signal processing to generate wireless signals complying with the 5G

Table 2. Terminology used in our work and their equivalents in 3rd Generation Partnership Project (3GPP) standards [1,2,4].

Our work	3GPP standards
User	User Equipment (UE)
Base station	gnodeB (gNB)
Core network	Access and Mobility Management Function (AMF), Session Management Function (SMF), Unified Data Management (UDM), and Policy Control Function (PCF)
Identity request	Non-Access Stratum (NAS) Identity Request

NR standard) and can detect and listen another legitimate user's transmission (passive radio receiving capability). In our threat model, the attacker knows the 5G protocols including the structure of control communication messages and preambles by Kerckhoff's principle. The attacker is implemented on the user side within the communication range to the base station; thus, there is no need to compromise the service provider infrastructure, e.g., the base station itself or a router between the base station and the core network. The attacker generates and transmits the preamble, initial wireless packets, and registration request to engage the base station and core network for Registration Request (from the base station's Broadcasting Message to the attacker user's Identity Request in Fig. 2). The injection attack triggers computing and networking loads to the 5G service provider infrastructure including the base station and the core network.

The Vulnerability. The vulnerability of our injection threat is from the initial (wireless) channel access setup forgoing any security protection. This is a part of the registration process and control communications before the authentication and the security mode setup, described in Sect. 2. While the attacker only engages the base station for the injection attack, the attack impacts the entire 5G networking service provider including the core network.

High-Feasibility Threat. Our threat is a high-risk threat since it involves a very low-barrier attack setup/feasibility as discussed in this section while providing great impact on the networking service provider. Our paper focuses on quantifying and measuring the injection attack impact. We focus on the per-injection attack impact analyses. An attacker model with greater constraints (and thus lower feasibility) and with potentially greater sophistication, e.g., those requiring real-time eavesdropping enabling spoofing, may have greater attack impacts, which analyses and studies we leave for future work.

4 Threat Impact Measurements and Analyses

In this section, we analyze the attack impact on the base station and core networking based on our cellular networking experimentation measurements.

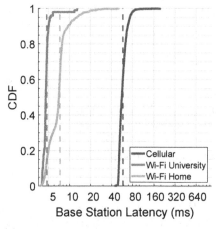
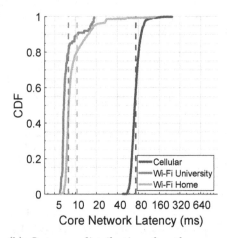

(a) Latency distribution for the base station. The confidence intervals with 95% certainty for base station are ±0.586ms, ±0.2234ms, and ±0.2202ms for Cellular, Wi-Fi University, and Wi-Fi Home latencies, respectively.

(b) Latency distribution for the core network. The confidence intervals with 95% certainty for core network are ±0.8486ms, ±0.578ms, and ±0.6669ms for Cellular, Wi-Fi University, and Wi-Fi Home latencies, respectively.

Fig. 3. The cumulative distribution function (CDF) for the initial base station/core network node associated with cellular network, university Wi-Fi, and home Wi-Fi. The dotted line shows the mean latencies for each platform.

4.1   Experiment Methodology

We experiment on real-world networking from the client perspective to estimate the injection attack impact, i.e., we take networking measurements on the client node networking with the 5G cellular service providers and estimate the attack impact based on the measurements. Because we experiment against real-world networking, we avoid DoS'ing or negatively impacting the service providers and only collect the reference measurements for our injection impact estimation. We select the top ten most used web domains [5] as of August 2021, and run the built-in *traceroute* shell command to identify every node in the route along with the link latencies. Our measurements are taken using a physical machine with Apple M1 chip, 8 core CPU (4 performance cores and 4 efficiency cores), 16 GBs (LPDDR4) of RAM running Python 3.8 for experimentation automation. While we focus our study on 5G cellular technology ("Cellular"), we do a comparison analyses with Wi-Fi and more specifically focusing on the university Wi-Fi networking infrastructure ("Wi-Fi University") and the residence Wi-Fi infrastructure ("Wi-Fi Home") since the university networking infrastructure and the residential networking exhibit different networking behavior and characteristics [11]. The actual service providers are anonymized on this manuscript because we are studying and simulating threats and our threat is applicable to 5G networking and not service-provider specific.

We run our experiment for the aforementioned networking cases and collect 1000 samples for each case. We identify the core network nodes from the networking service providers by cross-checking with the service provider IP range from Internet Assigned Numbers Authority [14] and then measure the number of hops and the latencies to reach those core network servers.

4.2 Networking Measurements for References

We compare the 5G networking with Wi-Fi networking, as Wi-Fi has existed longer and is more widely studied in networking. Cellular networking is generally with greater time costs than the Wi-Fi networking (which observation corroborates with [22]) and thus the threat impact is greater than the DoS impact on Wi-Fi. Figure 3a shows the distribution of latencies with regards to the base station. The base station mean latencies for cellular, university Wi-Fi, and home Wi-Fi are 58.45, 4.77, and 7.29 ms, respectively. Our results show that the Wi-Fi-based measurements have the lower latencies (12.25 and 8.02 times faster than cellular for university and home Wi-Fi, respectively), with university being 1.53 times faster than home Wi-Fi, on average.

Figure 3b shows the distribution of latencies with regards to the core network. The core network mean latencies for cellular, university Wi-Fi, and home Wi-Fi are 73.51, 7.93, and 10.54 ms, respectively. We find that the Wi-Fi-based measurements have the lowest latencies (9.27 and 6.97 times faster than cellular for university and home Wi-Fi, respectively) with university being 1.33 times faster than home Wi-Fi, on average.

Our estimation analysis in Sect. 4.3 focuses on the cellular measurements and we denote the number of communications between the user and base station (one interaction involving back-and-forth communications) as C_{BS} and the number of communications between the base station and core network as C_{CN}. We also denote the number of hops between the user and base station as H_{BS} while we use H_{CN} for the number of hops between the user and the core network. Finally, we use T_{BS} to denote the base station latency and T_{CN} to denote the core network latency. From our measurements, we have $H_{BS} = 1$ hop (typical since the user directly communicates with the base station via wireless communication link), $H_{CN} = 4$ hops, $T_{BS} = 58.45$ ms, $T_{CN} = 73.51$ ms.

4.3 Injection DoS Impact Estimation

We estimate the injection attack impact using our networking reference measurements in Sect. 4.2. More specifically, we quantify and estimate the injection attack impact in the number of communications (one back-and-forth communications regardless of whether it is a base station or a cloud-based core network), C, the number of hops, H, and latency, T. We compute C by calculating the total number of round-trip communications to the base station and core network, i.e., $C = C_{BS} + C_{CN}$. For example, from Fig. 2, the number of round-trip communications to the base station, C_{BS}, is 2 and to the core network, C_{CN}, is 1, giving $C = 3$ (until identity request). If the attacker chooses to send a response, it would

Table 3. Injection Attack Impact Estimations in the number of communications, C, the number of hops H, and latency T.

Identity response	C (communications)	H (hops)	T (ms)
Without	3	12	190.41
With	4	20	263.92

cost an additional round trip time. Similarly, the total number of hops involved in a round-trip communication is given as twice the product of the number of hops and the number of communications, $H = 2 \cdot (C_{BS} \cdot H_{BS} + C_{CN} \cdot H_{CN})$. The total latency is given as the product of the number of communications to the base station/core network and their respective latencies, $T = T_{BS} \cdot C_{BS} + T_{CN} \cdot C_{CN}$. Using the above equations, we quantify the injection impact. Table 3 computes the injection impact based on our reference measurements in Sect. 4.2 when an attacker does an injection without and with identity response. If the attacker injects a false identity response, it's going to cause an additional round-trip communication involving 8 more hops and taking $T = 263.92$ ms in total.

5   Related Work

In wireless and mobile security, previous research studied the control communication injection threats targeting the victim's availability. These research studies defended against DoS threats based on jamming the control communication channel itself because the control communication channel is publicly known [6,9,16], injecting false information on the wireless-channel-setting medium access control (MAC) communications for DoS [8,19], and injecting false information on the MAC feedback [23]. While our threat can classify as false/bogus control communication injection (more specifically, false requests for the initial channel requests), our work focuses on the emerging 5G networking as opposed to the more general wireless networking.

In 5G networking, similar to previous networking technologies, most of the well-known attacks like spoofing, sniffing, signaling, amplification are applicable [17,18,24]. The authors of [18] provide threat assessment and mitigation techniques for each of the control channels (broadcast, random access, uplink control, and downlink control) and data channels (uplink and downlink) of 5G. Their assessment shows that spoofing/jamming/sniffing attack efficiency on control channels is more effective. A signaling DoS attack on 3G/WiMax is presented in [17], where a 40-byte packet is sent to 24,000 mobile devices after every 5 s, which generates enough data to overload the used wireless infrastructure. The authors of [24] implement the signaling attack on 4G, classified the impact levels, and discussed possible countermeasures. The authors in [20] show that distributed DoS poses a big threat for 5G network slices. They also discuss the optimal placement for virtual network functions for guaranteed end-to-end delays. A strategic approach based on game theory is proposed in [21] to secure a 5G control plane from distributed DoS signaling attacks. It involves scaling up or increasing the resource assignment for virtual hosts based on the incoming signaling traffic.

6   Future Work and Potential Countermeasure Discussions

We discuss the future scope and potential countermeasures for strengthening the security of control communication. We intend to inform the 5G standard technologies of the injection threat in order to inform the research and developments for securing the networking availability. To achieve such a goal, we identify the following future research directions:

More Sophisticated Threats. Our threat model is of high feasibility merely requiring the 5G wireless communication capability, as described in Sect. 3. A more sophisticated threat may provide a stronger attack impact on the 5G service provider's availability but may impose greater constraints for the attacker setup.

Further Injection DoS Impact Analyses. Our work focuses on the per-injection impact analyses due to our experimental setup, including client-node implementation and experimentation against the real-world networking (Sect. 4.1), which can be used to estimate the attacker cost vs. impact for DoS involving multiple injections. Future research can therefore study the threat impact analyses of multiple injections, including flooding, DDoS, and their impacts on the victim's bandwidth and other networking resources. Analyzing the DoS impact on specific parts of the networking infrastructure, e.g., base station and a core network server functionality, can also identify the bottleneck vulnerabilities of the service provider infrastructure. Furthermore, there can be other threat impact metrics, including the wireless channel/medium-access-control resources (between the user and the base station) [8,17–19,23], host-based networking resources (similar to a TCP SYN Flood attacking the server connection table) [7,15,18,25], and the power resource (especially useful if considering a flying base station to provide emergency networking or future-generation networking with greater device/rate requirements for connectivity services) [10,12,13].

Building Security on the Base Station and Networking Edge. Security for 5G networking can be implemented at multiple levels. The base station or an edge server closer to the attacker user than the core network can build intelligence based on networking/sensed data to *detect* the injection threats to inform the attack. If such intelligence is real-time (i.e., occurring while the attack is ongoing), the base station can filter the attack traffic for *mitigation* so that the attack impact does not reach beyond the edge of the network. The base station on the networking edge can also build *prioritization* of the users, for example, based on token credentials from the previous session, so that the priority users demonstrating greater trustworthiness than others can still access the networking service (i.e., not being subjected to DoS). These approaches will mitigate the attack impact, thus reducing the security risk, and remain as a critical research direction to secure mobile/wireless networking.

5G Implementations. Our work studies the security based on the 5G NR standard and discover and exploit the vulnerability from the security being absent initially during the channel access setup before the authentication. A more systems approach based on the 5G system implementations and the collaborations with a real-world cellular service provider will facilitate more concrete analyses on the attack impact of the injection analyses. Such an approach will also enable the practicality analyses of the future security solutions for 5G networking.

Securing 6G and Next-Generation Networking. As observed from the mobile networking technology evolutions from 2G to 5G, mobile networking learn from its past generation of technologies to build the future-generation of networking technologies. Our research in securing the existing mobile networking technologies will contribute to building security in the next-generation technologies, such as 6G. Such research will enable the design of the security mechanisms along with the design of the 6G technology functionalities so that we can practice the security-by-design principle that would improve the security mechanisms' effectiveness and practicality beyond building security around the existing technologies. It will also drive and enable the security solution incorporation into the standards for wider deployment of the security implementations.

7 Conclusion

There is no security protection at the initial parts of the 5G control communications before the authentication and the security setup. The attacker can exploit this vulnerability to perform injection attack to consume the networking resources of the 5G service provider infrastructure. In this work, we study the 5G networking standardization and analyze the injection threat impact against the networking availability. We conduct networking measurements on real-world 5G and estimate the injection threat impacts in communications, hops, and latencies based on the networking measurements. We intend to inform the 5G standard technologies of the injection threat to encourage and facilitate the R&D in securing the networking availability. To that end, we include discussions for future work for greater security awareness and potential countermeasures, including the security mechanisms implemented at the base station or the networking edge.

Acknowledgement. This work was supported in part by Institute of Information & communications Technology Planning & Evaluation (IITP) grant funded by the Korea government (MSIT) (No.2021-0-00796, Research on Foundational Technologies for 6G Autonomous Security-by-Design to Guarantee Constant Quality of Security).

This material is also based upon work supported by the National Science Foundation under Grant No. 1922410.

References

1. 3GPP. TR 21.915: Release 15 Description; Summary of Rel-15 Work Items (2019). https://www.3gpp.org/release-15
2. 3GPP. TR 21.915: 5G; Procedures for the 5G System (2021). https://www.3gpp.org/specifications/specifications
3. 3GPP. TS 38.211: 5G; NR; Physical channels and modulation (2021). https://www.3gpp.org/specifications/specifications
4. 3GPP. TS 38.321: NR; Medium Access Control (MAC) protocol specification (2021). https://www.3gpp.org/specifications/specifications
5. Alexa: The top 500 sites on the web (2021). https://www.alexa.com/topsites
6. Arjoune, Y., Faruque, S.: Smart jamming attacks in 5G new radio: a review. In: 2020 10th Annual Computing and Communication Workshop and Conference (CCWC), pp. 1010–1015. IEEE (2020)
7. Bogdanoski, M., Suminoski, T., Risteski, A.: Analysis of the SYN flood dos attack. Int. J. Comput. Netw. Inf. Secur. (IJCNIS) 5(8), 1–11 (2013)
8. Chang, S.Y., Hu, Y.C.: SecureMAC: securing wireless medium access control against insider denial-of-service attacks. IEEE Trans. Mob. Comput. 16(12), 3527–3540 (2017). https://doi.org/10.1109/TMC.2017.2693990
9. Chang, S.Y., Hu, Y.C., Laurenti, N.: SimpleMAC: a jamming-resilient MAC-layer protocol for wireless channel coordination. In: Proceedings of the 18th Annual International Conference on Mobile Computing and Networking, Mobicom'12, pp. 77–88. Association for Computing Machinery, New York (2012). https://doi.org/10.1145/2348543.2348556
10. Chang, S.Y., Kumar, S.L.S., Hu, Y.C., Park, Y.: Power-positive networking: wireless-charging-based networking to protect energy against battery dos attacks. ACM Trans. Sen. Netw. 15(3), 1–25 (2019). https://doi.org/10.1145/3317686
11. Chang, S.Y., Park, Y., Kengalahalli, N.V., Zhou, X.: Query-crafting DoS threats against internet DNS. In: 2020 IEEE Conference on Communications and Network Security (CNS), pp. 1–9 (2020). https://doi.org/10.1109/CNS48642.2020.9162166
12. Desnitsky, V., Rudavin, N., Kotenko, I.: Modeling and evaluation of battery depletion attacks on unmanned aerial vehicles in crisis management systems. In: Kotenko, I., Badica, C., Desnitsky, V., El Baz, D., Ivanovic, M. (eds.) IDC 2019. SCI, vol. 868, pp. 323–332. Springer, Cham (2020). https://doi.org/10.1007/978-3-030-32258-8_38
13. Halperin, D., et al.: Pacemakers and implantable cardiac defibrillators: software radio attacks and zero-power defenses. In: 2008 IEEE Symposium on Security and Privacy (SP 2008), pp. 129–142 (2008). https://doi.org/10.1109/SP.2008.31
14. IANA: Internet Assigned Numbers Authority. https://www.iana.org/
15. Kolahi, S.S., Alghalbi, A.A., Alotaibi, A.F., Ahmed, S.S., Lad, D.: Performance comparison of defense mechanisms against TCP SYN flood DDoS attack. In: 2014 6th International Congress on Ultra Modern Telecommunications and Control Systems and Workshops (ICUMT), pp. 143–147. IEEE (2014)
16. Lazos, L., Liu, S., Krunz, M.: Mitigating control-channel jamming attacks in multi-channel ad hoc networks. In: Proceedings of the Second ACM Conference on Wireless Network Security, WiSec'09, pp. 169–180. Association for Computing Machinery, New York (2009). https://doi.org/10.1145/1514274.1514299

17. Lee, P.P., Bu, T., Woo, T.: On the detection of signaling DoS attacks on 3G/WiMax wireless networks. Comput. Netw. **53**(15), 2601–2616 (2009)

18. Lichtman, M., Rao, R., Marojevic, V., Reed, J., Jover, R.P.: 5G NR jamming, spoofing, and sniffing: threat assessment and mitigation. In: 2018 IEEE International Conference on Communications Workshops (ICC Workshops), pp. 1–6. IEEE (2018)

19. Negi, R., Rajeswaran, A.: DoS analysis of reservation based MAC protocols. In: IEEE International Conference on Communications, ICC 2005, vol. 5, pp. 3632–3636 (2005). https://doi.org/10.1109/ICC.2005.1495094

20. Sattar, D., Matrawy, A.: Towards secure slicing: using slice isolation to mitigate DDoS attacks on 5G core network slices. In: 2019 IEEE Conference on Communications and Network Security (CNS), pp. 82–90. IEEE (2019)

21. Silva, R.S., et al.: REPEL: a strategic approach for defending 5G control plane from DDoS signalling attacks. IEEE Trans. Netw. Serv. Manag. **18**(3), 3231–3243 (2020)

22. Sommers, J., Barford, P.: Cell vs. WiFi: on the performance of metro area mobile connections. In: Proceedings of the 2012 Internet Measurement Conference, pp. 301–314 (2012)

23. Tung, Y.C., Han, S., Chen, D., Shin, K.G.: Vulnerability and protection of channel state information in multiuser MIMO networks. In: Proceedings of the 2014 ACM SIGSAC Conference on Computer and Communications Security, CCS'14, pp. 775–786. Association for Computing Machinery, New York (2014). https://doi.org/10.1145/2660267.2660272

24. Yu, C., Chen, S.: On effects of mobility management signalling based DoS attacks against LTE terminals. In: 2019 IEEE 38th International Performance Computing and Communications Conference (IPCCC), pp. 1–8. IEEE (2019)

25. Zhang, T., Lee, R.B.: Host-based DoS attacks and defense in the cloud. In: Proceedings of the Hardware and Architectural Support for Security and Privacy, pp. 1–8 (2017)

Author Index

Printed in the United States
by Baker & Taylor Publisher Services